Lecture Notes in Mathematics 2055

For further volumes:
http://www.springer.com/series/304

Sungbok Hong • John Kalliongis
Darryl McCullough • J. Hyam Rubinstein

Diffeomorphisms of Elliptic 3-Manifolds

Springer

Sungbok Hong
Korea University
Department of Mathematics
Seoul, Korea

John Kalliongis
Saint Louis University
Department of Mathematics
and Computer Science
St. Louis, MO, USA

Darryl McCullough
University of Oklahoma
Department of Mathematics
Norman, OK, USA

J. Hyam Rubinstein
University of Melbourne
Department of Mathematics
Melbourne, Victoria, Australia

ISBN 978-3-642-31563-3 ISBN 978-3-642-31564-0 (eBook)
DOI 10.1007/978-3-642-31564-0
Springer Heidelberg New York Dordrecht London

Lecture Notes in Mathematics ISSN print edition: 0075-8434
ISSN electronic edition: 1617-9692

Library of Congress Control Number: 2012945525

Mathematics Subject Classification (2010): 57M99, 57S10, 58D05, 58D29

Printed on acid-free paper

Springer is part of Springer Science+Business Media (www.springer.com)

Preface

This work is ultimately directed at understanding the diffeomorphism groups of elliptic three-manifolds—those closed three-manifolds that admit a Riemannian metric of constant positive curvature. The main results concern the Smale Conjecture. The original Smale Conjecture, proven by A. Hatcher [24], asserts that if M is the 3-sphere with the standard constant curvature metric, the inclusion $\mathrm{Isom}(M) \to \mathrm{Diff}(M)$ from the isometry group to the diffeomorphism group is a homotopy equivalence. The *Generalized Smale Conjecture* (henceforth just called the Smale Conjecture) asserts this whenever M is an elliptic three-manifold.

Here are our main results:

1. The Smale Conjecture holds for elliptic three-manifolds containing geometrically incompressible Klein bottles (Theorem 1.2). These include all quaternionic and prism manifolds.
2. The Smale Conjecture holds for all lens spaces $L(m, q)$ with $m \geq 3$ (Theorem 1.3).

Many of the cases in Theorem 1.2 were proven a number of years ago by N. Ivanov [33–36] (see Sect. 1.2).

Some of our other results concern the groups of diffeomorphisms $\mathrm{Diff}(\Sigma)$ and fiber-preserving diffeomorphisms $\mathrm{Diff}_f(\Sigma)$ of a Seifert-fibered Haken three-manifold Σ and the coset space $\mathrm{Diff}(\Sigma)/\mathrm{Diff}_f(\Sigma)$, which is called the space of Seifert fiberings (equivalent to the given fibering) of Σ.

3. Apart from a small list of known exceptions, $\mathrm{Diff}_f(\Sigma) \to \mathrm{Diff}(\Sigma)$ is a homotopy equivalence (Theorem 3.15).
4. The space of Seifert fiberings of Σ has contractible components (Theorem 3.14) and apart from a small list of known exceptions, it is contractible (Theorem 3.15).

These may be already accepted as part of the overall three-dimensional landscape, but we are unable to find any serious treatment of them. And we have found that the development of the necessary tools and their application to the three-dimensional context goes well beyond a routine exercise.

Table 1 Status of the Smale conjecture

Case	SC proven?
S^3	Hatcher [24]
$\mathbb{RP}^3$	
Lens spaces	Chapter 5
Prism and quaternionic manifolds	Ivanov [33–36] and Chap. 4
Tetrahedral manifolds	
Octahedral manifolds	
Icosahedral manifolds	

This manuscript includes work done more than 20 years ago, as well as work recently completed. In the mid-1980s, two of the authors (DM and JHR) sketched an argument proving the Smale Conjecture for the three-manifolds that contain one-sided Klein bottles (other than the lens space $L(4, 1)$). That method, which ultimately became Chap. 4, underwent a long evolution as various additions were made to fill in technical details.

The case of one-sided Klein bottles includes some lens spaces—those of the form $L(4n, 2n - 1)$ for $n \geq 2$. But for the general lens space case, a different approach using Heegaard tori was developed by SH and DM starting around 2000. It is based on a powerful methodology developed by JHR and M. Scharlemann [58]. It turned out that JHR was working on the Smale Conjecture for lens spaces along exactly the same lines as SH and DM, so the efforts were combined in the work that became Chap. 5.

One more case of the Smale Conjecture may be accessible to existing techniques. It seems likely that A. Hatcher's approach to the S^3 case in [24] would also serve for $\mathbb{RP}^3$, but this has yet to be carried out.

In summary, this is where the Smale Conjecture now stands (Table 1).

Our work on the Smale Conjecture requires some basic theory about spaces of mappings of smooth manifolds, such as the fact that diffeomorphism groups of compact manifolds and spaces of embeddings of submanifolds have the homotopy type of CW-complexes, a result originally proven by R. Palais. This theory is well known to global analysts and others, but not to many low-dimensional topologists. Also, most sources do not discuss the case of manifolds with boundary, and we know of no existing treatment of the case of fiber-preserving diffeomorphisms and embeddings, which is the context of much of our technical work. For this reason, we have included a fair dose of foundational material on diffeomorphism groups in Chap. 2, which includes the case of manifolds with boundary, with the additional boundary control that we will need.

A more serious gap in the literature is the absence of versions of the fundamental restriction fibration theorems of Palais and Cerf in the context of fibered (and Seifert-fibered) manifolds. These extensions of the well-known theory require some new ideas, which were developed by JK and DM and form most of Chap. 3. We work in a class of singular fiberings large enough to include all Seifert fiberings of three-manifolds, except some fiberings of lens spaces. These results are heavily used in our work in Chaps. 4 and 5. Our results on fiber-preserving diffeomorphisms and the space of fibered structures of a Seifert-fibered Haken three-manifold are applications of this work, and they also appear in Chap. 3.

Much of our work in this text is unusually detailed and technical. In considerable part, this not only arises from its inherent complication, but it also reflects the fact that over the years we have filled in many arguments in response to recommendations from various readers. Unfortunately, one reader's "too sketchy" can be another's "too much elaboration of well-known facts," and personally we find some of the current exposition to be somewhat too long and too detailed. To provide an alternative, we have included Sects. 4.2 and 5.1, which are overviews of the proofs of the main results. In the actual proofs, we trust that each reader will simply accept the "obvious" parts and focus on the "nontrivial" parts, whichever they may be.

We have made the text self-contained, when possible, and sought useful references when not. We do assume that the reader is comfortable with basic topology and differential topology of manifolds, group actions on manifolds, Riemannian metrics, fibrations, and so on. We freely use classical three-manifold topology, such as I-bundles and Seifert-fibered structures and two-dimensional orbifolds, the Jaco–Shalen–Johannson decomposition, the results of Waldhausen, and hyperbolic three-manifolds, as well as major developments such as the results (but not the methods) of Perelman. We rather freely use facts about spaces of isometries and diffeomorphisms of surfaces and commonly encountered three-manifolds. Here, familiarity with papers of A. Hatcher such as [22, 23] would be very helpful. In the realm of infinite-dimensional topology, we use some basics about Fréchet manifolds and some standard theorems, see Sect. 2.1 for a discussion. Additionally we draw on the theory of singularities, modestly in Chap. 4 and in quite a bit more depth in Chap. 5. Both chapters make heavy use of the parameterized methods in the aforementioned papers of Hatcher, and in the latter chapter, familiarity with the Rubinstein–Scharlemann graphic [58] will be very helpful.

The authors are grateful to many sources of support during the lengthy preparation of this work. These include the Australian Research Council, the Korea Research Foundation, the Basic Science Research Center of Korea University, Saint Louis University, the U.S. National Science Foundation, the Mathematical Sciences Research Institute, the University of Oklahoma Vice President for Research, and the University of Oklahoma College of Arts and Sciences. We also thank the referees of versions of this work for occasional corrections and numerous helpful suggestions, as well as the editors and staff at Springer for their work to produce this final version.

Contents

Chapter 1
Elliptic Three-Manifolds and the Smale Conjecture

As noted in the Preface, the Smale Conjecture is the assertion that the inclusion $\mathrm{Isom}(M) \to \mathrm{Diff}(M)$ is a homotopy equivalence whenever M is an elliptic three-manifold, that is, a three-manifold with a Riemannian metric of constant positive curvature (which may be assumed to be 1). The Geometrization Conjecture, now proven by Perelman, shows that all closed three-manifolds with finite fundamental group are elliptic.

In this chapter, we will first review elliptic three-manifolds and their isometry groups. In the second section, we will state our main results on the Smale Conjecture, and provide some historical context. In the final two sections, we discuss isometries of nonelliptic three-manifolds, and address the possibility of applying Perelman's methods to the Smale Conjecture.

1.1 Elliptic Three-Manifolds and Their Isometries

The elliptic three-manifolds were completely classified long ago. They are exactly the three-manifolds whose universal cover can be uniformized as the unit sphere S^3 in $\mathbb{R}^4$ so that $\pi_1(M)$ acts freely as a subgroup of $\mathrm{Isom}_+(S^3) = \mathrm{SO}(4)$. The subgroups of SO(4) that act freely were first determined by Hopf and Seifert–Threlfall, and reformulated using quaternions by Hattori. References include [74] (pp. 226–227), [49] (pp. 103–113), [60] (pp. 449–457), [46, 59].

The isometry groups of elliptic three-manifolds have also been known for a long time, and are topologically rather simple: they are compact Lie groups of dimension at most 6. A detailed calculation of the isometry groups of elliptic three-manifolds was given in [46], and in this section we will recall the resulting groups.

To set notation, recall that there is a well-known twofold covering $S^3 \to \mathrm{SO}(3)$, which is a homomorphism when S^3 is regarded as the group of unit quaternions (see Sect. 4.3 for a fuller discussion). The elements of SO(3) that preserve a given axis, say the z-axis, form the orthogonal subgroup O(2). We will denote

S. Hong et al., *Diffeomorphisms of Elliptic 3-Manifolds*, Lecture Notes in Mathematics 2055, DOI 10.1007/978-3-642-31564-0_1,

Table 1.1 Isometry groups of $M = S^3/G$ $(m > 2, n > 1)$

G	M	Isom(M)	$\mathscr{I}(M)$
Q_8	Quaternionic	$SO(3) \times S_3$	S_3
$Q_8 \times C_n$	Quaternionic	$O(2) \times S_3$	$C_2 \times S_3$
D^*_{4m}	Prism	$SO(3) \times C_2$	C_2
$D^*_{4m} \times C_n$	Prism	$O(2) \times C_2$	$C_2 \times C_2$
Index 2 diagonal	Prism	$O(2) \times C_2$	$C_2 \times C_2$
T^*_{24}	Tetrahedral	$SO(3) \times C_2$	C_2
$T^*_{24} \times C_n$	Tetrahedral	$O(2) \times C_2$	$C_2 \times C_2$
Index 3 diagonal	Tetrahedral	$O(2)$	C_2
O^*_{48}	Octahedral	$SO(3)$	$\{1\}$
$O^*_{48} \times C_n$	Octahedral	$O(2)$	C_2
I^*_{120}	Icosahedral	$SO(3)$	$\{1\}$
$I^*_{120} \times C_n$	Icosahedral	$O(2)$	C_2

by $O(2)^*$ the inverse image in S^3 of $O(2)$. When H_1 and H_2 are groups, each containing -1 as a central involution, the quotient $(H_1 \times H_2)/\langle(-1,-1)\rangle$ is denoted by $H_1 \tilde{\times} H_2$. In particular, SO(4) itself is $S^3 \tilde{\times} S^3$, and contains the subgroups $S^1 \tilde{\times} S^3$, $O(2)^* \tilde{\times} O(2)^*$, and $S^1 \tilde{\times} S^1$. The latter is isomorphic to $S^1 \times S^1$, but it is sometimes useful to distinguish between them. Finally, $\mathrm{Dih}(S^1 \times S^1)$ is the semidirect product $(S^1 \times S^1) \circ C_2$, where C_2 acts by complex conjugation in both factors.

There are twofold covering homomorphisms

$$O(2)^* \times O(2)^* \to O(2)^* \tilde{\times} O(2)^* \to O(2) \times O(2) \to O(2) \tilde{\times} O(2)\,.$$

Each of these groups is diffeomorphic to four disjoint copies of the torus, but they are pairwise nonisomorphic. Indeed, they are easily distinguished by examining their subsets of order two elements. Similarly, $S^1 \times S^3$ and $S^1 \tilde{\times} S^3$ are diffeomorphic, but nonisomorphic.

Table 1.1 gives the isometry groups of the elliptic three-manifolds with noncyclic fundamental group. The first column, G, indicates the fundamental group of M, where C_m denotes a cyclic group of order m, and D^*_{4m}, T^*_{24}, O^*_{48}, and I^*_{120} are the binary dihedral, tetrahedral, octahedral, and icosahedral groups of the indicated orders. The groups called index 2 and index 3 diagonal are certain subgroups of $D^*_{4m} \times C_{4m}$ and $T^*_{24} \times C_{6n}$ respectively. The last two columns give the full isometry group $\mathrm{Isom}(M)$, and the group $\mathscr{I}(M)$ of path components of $\mathrm{Isom}(M)$.

Table 1.2 gives the isometry groups of the elliptic three-manifolds with cyclic fundamental group. These are the 3-sphere $L(1,0)$, real projective space $L(2,1)$, and the lens spaces $L(m,q)$ with $m \geq 3$.

Section 4.3 contains the detailed calculation of $\mathrm{isom}(M)$, the connected component of id_M in $\mathrm{Isom}(M)$, for the elliptic three-manifolds that contain one-sided incompressible Klein bottles (the quaternionic and prism manifolds, and the lens spaces of the form $L(4n, 2n-1)$), since the notation and some of the mechanics of this calculation are needed for the arguments in Chap. 4.

Table 1.2 Isometry groups of $L(m,q)$

m, q	$\mathrm{Isom}(L(m,q))$	$\mathscr{I}(L(m,q))$
$m = 1$ ($L(1,0) = S^3$)	$O(4)$	C_2
$m = 2$ ($L(2,1) = \mathbb{RP}^3$)	$(SO(3) \times SO(3)) \circ C_2$	C_2
$m > 2$, m odd, $q = 1$	$O(2)^* \widetilde{\times} S^3$	C_2
$m > 2$, m even, $q = 1$	$O(2) \times SO(3)$	C_2
$m > 2, 1 < q < m/2, q^2 \not\equiv \pm 1 \bmod m$	$\mathrm{Dih}(S^1 \times S^1)$	C_2
$m > 2, 1 < q < m/2, q^2 \equiv -1 \bmod m$	$(S^1 \widetilde{\times} S^1) \circ C_4$	C_4
$m > 2, 1 < q < m/2, q^2 \equiv 1 \bmod m$, $\gcd(m, q+1)\gcd(m, q-1) = m$	$O(2) \widetilde{\times} O(2)$	$C_2 \times C_2$
$m > 2, 1 < q < m/2, q^2 \equiv 1 \bmod m$, $\gcd(m, q+1)\gcd(m, q-1) = 2m$	$O(2) \times O(2)$	$C_2 \times C_2$

1.2 The Smale Conjecture

Smale [64] proved that for the standard round 2-sphere S^2, the inclusion of the isometry group $O(3)$ into the diffeomorphism group $\mathrm{Diff}(S^2)$ is a homotopy equivalence. He conjectured that the analogous result holds true for the 3-sphere, that is, that $O(4) \to \mathrm{Diff}(S^3)$ is a homotopy equivalence. Cerf [11] proved that the inclusion induces a bijection on path components, and the full conjecture was proven by Hatcher [24].

A weak form of the (generalized) Smale Conjecture is known. In [46], the calculations of $\mathrm{Isom}(M)$ for elliptic three-manifolds are combined with results on mapping class groups of many authors, including [2, 5, 6, 56, 57], to obtain the following statement:

Theorem 1.1. *Let M be an elliptic three-manifold. Then the inclusion of* $\mathrm{Isom}(M)$ *into* $\mathrm{Diff}(M)$ *is a bijection on path components.*

This can be called the "π_0-part" of the Smale Conjecture. By virtue of this result, to prove the Smale Conjecture for any elliptic three-manifold, it is sufficient to prove that the inclusion $\mathrm{isom}(M) \to \mathrm{diff}(M)$ of the connected components of the identity map in $\mathrm{Isom}(M)$ and $\mathrm{Diff}(M)$ is a homotopy equivalence.

The earliest work on the Smale Conjecture was by N. Ivanov. Certain elliptic three-manifolds contain one-sided geometrically incompressible Klein bottles. Fixing such a Klein bottle K_0, called the base Klein bottle, the remainder of the three-manifold is an open solid torus, and (up to isotopy) there are two Seifert fiberings, one for which the Klein bottle is fibered by nonsingular fibers (the "meridional" fibering), and one for which it contains two exceptional fibers of type $(2,1)$ (the "longitudinal" fibering). As will be detailed in Sect. 4.1 below, the manifolds then fall into four types:

(I) Those for which neither the meridional nor the longitudinal fibering is nonsingular on the complement of K_0.
(II) Those for which only the longitudinal fibering is nonsingular on the complement of K_0. These are the lens spaces $L(4n, 2n-1)$, $n \geq 2$.

(III) Those for which only the meridional fibering is nonsingular on the complement of K_0.
(IV) The lens space $L(4,1)$, for which both the meridional and longitudinal fiberings are nonsingular on the complement of K_0.

Cases I and III are the quaternionic and prism manifolds.

Ivanov announced the Smale Conjecture for Cases I and II in [33,34], and gave a detailed proof for Case I in [35,36]. One of our main theorems extends those results to all cases:

Theorem 1.2 (Smale Conjecture for elliptic three-manifolds containing incompressible Klein bottles). *Let M be an elliptic three-manifold containing a geometrically incompressible Klein bottle. Then* $\mathrm{Isom}(M) \to \mathrm{Diff}(M)$ *is a homotopy equivalence.*

Theorem 1.2 is proven in Chap. 4, except for the case of $L(4,1)$, which is proven in Chap. 5.

Our second main result concerns lens spaces, which for us refers only to the lens spaces $L(m,q)$ with $m \geq 3$:

Theorem 1.3 (Smale Conjecture for lens spaces). *For any lens space L, the inclusion* $\mathrm{Isom}(L) \to \mathrm{Diff}(L)$ *is a homotopy equivalence.*

One consequence of the Smale Conjecture is the determination of the homeomorphism type of $\mathrm{Diff}(M)$. Recall that a Fréchet space is a locally convex complete metrizable linear space. In Sect. 2.1, we will review the fact that if M is a closed smooth manifold, then with the C^∞-topology, $\mathrm{Diff}(M)$ is a separable infinite-dimensional manifold locally modeled on the Fréchet space of smooth vector fields on M. By the Anderson–Kadec Theorem [4, Corollary VI.5.2], every infinite-dimensional separable Fréchet space is homeomorphic to $\mathbb{R}^\infty$, the countable product of lines. A theorem of Henderson and Schori ([4, Theorem IX.7.3], originally announced in [28]) shows that if Y is any locally convex space with Y homeomorphic to Y^∞, then manifolds locally modeled on Y are homeomorphic whenever they have the same homotopy type. Applying this with $Y = \mathbb{R}^\infty$, our main theorems give immediately the homeomorphism type of $\mathrm{Diff}(M)$:

Corollary 1.1. *Let M be an elliptic three-manifold which either contains an incompressible Klein bottle or is a lens space $L(m,q)$ with $m \geq 3$. Then* $\mathrm{Diff}(M)$ *is homeomorphic to* $\mathrm{Isom}(M) \times \mathbb{R}^\infty$.

Combining this with the calculations of $\mathrm{Isom}(M)$ in Table 1.1 gives the following homeomorphism classification of $\mathrm{Diff}(M)$, in which P_n denotes the discrete space with n points:

Corollary 1.2. *Let M be an elliptic three-manifold, not a lens space, containing an incompressible Klein bottle.*

1. *If M is the quaternionic manifold with fundamental group $Q_8 = D_8^*$, then* $\mathrm{Diff}(M) \approx P_6 \times \mathrm{SO}(3) \times \mathbb{R}^\infty$.

2. *If M is a quaternionic manifold with fundamental group $Q_8 \times C_n$, $n > 2$, then* $\mathrm{Diff}(M) \approx P_{12} \times S^1 \times \mathbb{R}^\infty$.
3. *If M is a prism manifold with fundamental group D^*_{4m}, $m \geq 3$, then* $\mathrm{Diff}(M) \approx P_2 \times \mathrm{SO}(3) \times \mathbb{R}^\infty$.
4. *If M is any other prism manifold, then* $\mathrm{Diff}(M) \approx P_4 \times S^1 \times \mathbb{R}^\infty$.

As above, using Table 1.2, we obtain a complete classification of $\mathrm{Diff}(L)$ for lens spaces into four homeomorphism types:

Corollary 1.3. *For a lens space $L(m,q)$ with $m \geq 3$, the homeomorphism type of* $\mathrm{Diff}(L)$ *is as follows:*

1. *For m odd,* $\mathrm{Diff}(L(m,1)) \approx P_2 \times S^1 \times S^3 \times \mathbb{R}^\infty$.
2. *For m even,* $\mathrm{Diff}(L(m,1)) \approx P_2 \times S^1 \times \mathrm{SO}(3) \times \mathbb{R}^\infty$.
3. *For $q > 1$ and $q^2 \not\equiv \pm 1 \pmod m$,* $\mathrm{Diff}(L(m,q)) \approx P_2 \times S^1 \times S^1 \times \mathbb{R}^\infty$.
4. *For $q > 1$ and $q^2 \equiv \pm 1 \pmod m$,* $\mathrm{Diff}(L(m,q)) \approx P_4 \times S^1 \times S^1 \times \mathbb{R}^\infty$.

We remark that the homeomorphism classification is quite different from the isomorphism classification. In fact, for *any* smooth manifold, the *isomorphism type* of $\mathrm{Diff}(M)$ determines M. That is, an abstract isomorphism between the diffeomorphism groups of two differentiable manifolds must be induced by a diffeomorphism between the manifolds [3, 13, 66].

The Smale Conjecture has some other applications, beyond the problem of understanding $\mathrm{Diff}(M)$. Ivanov's results were used in [43] to construct examples of homeomorphisms of reducible three-manifolds that are homotopic but not isotopic. Our results show that the construction applies to a larger class of three-manifolds. In [55], Theorem 1.2 was applied to the classification problem for three-manifolds which have metrics of positive Ricci curvature and universal cover S^3.

The Smale Conjecture has attracted the interest of physicists studying the theory of quantum gravity. Certain physical configuration spaces can be realized as the quotient space of a principal $\mathrm{Diff}_1(M, x_0)$-bundle with contractible total space, where $\mathrm{Diff}_1(M, x_0)$ denotes the subgroup of $\mathrm{Diff}(M, x_0)$ that induce the identity on the tangent space to M at x_0. (This group is homotopy equivalent to $\mathrm{Diff}(M \# D^3 \text{ rel } \partial D^3)$.) Consequently the loop space of the configuration space is weakly homotopy equivalent to $\mathrm{Diff}_1(M, x_0)$. Physical significance of $\pi_0(\mathrm{Diff}(M))$ for quantum gravity was first pointed out in [14]. See also [1, 18, 30, 65, 73]. The physical significance of some higher homotopy groups of $\mathrm{Diff}(M)$ was examined by Giulini [17].

1.3 The Weak Smale Conjecture

For an arbitrary three-manifold M, we may say that M satisfies the Smale Conjecture if $\mathrm{Isom}(M) \to \mathrm{Diff}(M)$ is a homotopy equivalence for a Riemannian metric on M of maximal symmetry (that is, one for which the Lie group $\mathrm{Isom}(M)$ has maximal dimension and maximal number of components). In general, however, the SC does not extend beyond the elliptic case. The three-torus T^3 provides a simple example: $\mathrm{Diff}(T^3)$ has infinitely many components (since taking the induced outer automorphism on $\pi_1(T^3)$ defines a continuous surjection from $\mathrm{Diff}(T^3)$ onto $\mathrm{GL}(3,\mathbb{Z})$), but $\mathrm{Isom}(T^3)$ is a compact Lie group so has only finitely many components. In this example, however, the inclusion $\mathrm{isom}(M) \to \mathrm{diff}(M)$ of the connected components of the identity map in $\mathrm{Isom}(M)$ and $\mathrm{Diff}(M)$ is a homotopy equivalence. This and other examples motivate us to define the *Weak Smale Conjecture* (WSC) for M to be the assertion that the inclusion $\mathrm{isom}(M) \to \mathrm{diff}(M)$ is a homotopy equivalence. Note that the SC for M is equivalent to the assertion that $\mathrm{Isom}(M) \to \mathrm{Diff}(M)$ is a bijection on path components (the "π_0-part" of the conjecture) and the WSC, a fact used in the previous section to reduce the SC for elliptic three-manifolds to the WSC.

The WSC holds in some important cases, such as T^3, and the SC even extends for some classes of nonelliptic three-manifolds. In the remainder of this section, we will survey what is currently known for the nonelliptic closed orientable cases.

For closed Haken three-manifolds, $\mathrm{isom}(M)$ is $(S^1)^k$, where k is the rank of the center of $\pi_1(M)$. Explicitly, k is 3 when $M = T^3$, 1 for Seifert-fibered Haken three-manifolds with orientable quotient orbifold, and 0 otherwise. Work of Hatcher [22] and Ivanov [31, 32] shows that $\mathrm{isom}(M) \to \mathrm{diff}(M)$ is a homotopy equivalence, that is, the WSC (in [22], the results are stated for PL homeomorphisms, but the Smale Conjecture for S^3 extends the results to the smooth category).

Using his "insulator" methodology, 'Gabai [15] proved that the components of $\mathrm{Diff}(M)$ are contractible for all hyperbolic three-manifolds. He deduced the SC for these manifolds, showing in fact that both $\mathrm{Isom}(M) \to \mathrm{Diff}(M)$ and $\mathrm{Diff}(M) \to \mathrm{Out}(\pi_1(M))$ are homotopy equivalences for finite-volume hyperbolic three-manifolds (for hyperbolic three-manifolds that are also Haken, this was already known by Mostow Rigidity, Waldhausen's Theorem, and the work of Hatcher and Ivanov already discussed). The same statements have now been proven by Soma and the third author [47] when M has an $\mathbb{H}^2 \times \mathbb{R}$ or $\widetilde{\mathrm{SL}_2}(\mathbb{R})$ geometry and its (unique, up to isotopy) Seifert-fibered structure has base orbifold the 2-sphere with three cone points. This is expected to hold for the Nil geometry as well.

As for the non-irreducible case, Hatcher [23, 25] proved that $\mathrm{Diff}(S^2 \times S^1)$ is homotopy equivalent to $\mathrm{O}(2) \times \mathrm{O}(3) \times \Omega\,\mathrm{O}(3)$, where $\Omega\,\mathrm{O}(3)$ is the space of loops in O(3). In this case, the product metric is maximally symmetric and $\mathrm{Isom}(S^2 \times S^1)$ is diffeomorphic to $\mathrm{O}(2) \times \mathrm{O}(3)$, but the "rotation" components involving nontrivial elements of $\Omega\,\mathrm{O}(3)$ are not isotopic to isometries. The latter is geometrically obvious since no such element can preserve the geodesics of the form $\{x\} \times S^1$.

For the remaining non-irreducible three-manifolds, the WSC is known to fail in most cases. The second and third authors [39] proved that when M has at least three nonsimply connected prime summands, or one $S^2 \times S^1$-summand and one other prime summand with infinite fundamental group, $\pi_1(\mathrm{diff}(M))$ is not finitely generated, so the WSC fails drastically. When M is a connected sum $(S^2 \times S^1)\#P$ with $\pi_1(P)$ finite, the WSC fails at least when $\pi_1(P)$ has order more than 2. For by [39], $\pi_1(\text{diff}(M))$ has a free abelian summand of rank $n-1$, where n is the order of $\pi_1(P)$. On the other hand, $\pi_1(\text{Isom}(M))$ has rank at most 1. This can be seen using the fibration $\text{Isom}(M) \to M$ with fiber $\text{Isom}(M, x_0)$ the isometries preserving a basepoint of M. In the associated exact sequence, $\pi_1(\text{Isom}(M)) \to \pi_1(M)$ is the trivial homomorphism, since the trace of any isotopy from the identity to the identity is a central element of the fundamental group, and $\pi_1(M)$ is a nontrivial free product so is centerless. Therefore $\pi_1(\text{Isom}(M, x_0)) \to \pi_1(\text{Isom}(M))$ is surjective. Now $\text{Isom}(M, x_0)$ is a Lie subgroup of the isometries of the unit tangent 2-sphere of M at x_0, and so the connected component of the identity is a connected subgroup of SO(3) and can only be either trivial, S^1, or SO(3) itself (actually, the latter case cannot occur, since the action of $\text{Isom}(M, x_0)$ on M lifts to an action with fixed point on the Freudenthal endpoint compactification of the universal cover of M, which is S^3. The fixed point set of this action contains the Cantor set of endpoints, so has dimension at least 1).

1.4 Perelman's Methods

It is natural to ask whether the Smale Conjecture can be proven using the methodology that G. Perelman developed to prove the Geometrization Conjecture. The Smale Conjecture would follow if there were a flow retracting the space $\mathscr{R}$ of all Riemannian metrics on an elliptic three-manifold M to the subspace $\mathscr{R}_c$ of metrics of constant positive curvature. Here is why this is so. First, note that by rescaling, $\mathscr{R}_c$ deformation retracts to the subspace $\mathscr{R}_1$ of metrics of constant curvature 1. Now, $\text{Diff}(M)$ acts by pullback on $\mathscr{R}_1$; this action is transitive (given two constant curvature metrics on M, the developing map gives a diffeomorphism which is an isometry between the lifted metrics on the universal cover, and since the action of $\pi_1(M)$ is known to be unique up to conjugation by an isometry, this diffeomorphism can be composed with some isometry to make it equivariant) and the stabilizer of each point is a subgroup conjugate to $\text{Isom}(M)$, so $\mathscr{R}_1$ may be identified with the coset space $\text{Isom}(M)\backslash \text{Diff}(M)$. On the other hand, $\mathscr{R}$ is contractible (M is parallelizable and one can use a Gram–Schmidt orthonormalization process). So the existence of a flow retracting $\mathscr{R}$ to $\mathscr{R}_c$ would imply that $\text{Isom}(M)\backslash \text{Diff}(M)$ is contractible, which is equivalent to the Smale Conjecture. Finding a flow that retracts $\mathscr{R}$ to $\mathscr{R}_c$ is, of course, the rough idea of the Hamilton–Perelman program. At the present time, however, we do not see any way to carry this out, due to the formation of singularities and the requisite surgery of necks, and we are unaware of any progress in this direction.

Chapter 2
Diffeomorphisms and Embeddings of Manifolds

This chapter contains foundational material on spaces of diffeomorphisms and embeddings. Such spaces are known to be Fréchet manifolds, separable when the manifolds involved are compact. We will need versions of these and related facts for manifolds with boundary, and also in the context of fiber-preserving diffeomorphisms and maps. For the latter, a new (to us, at least) idea is required—the aligned exponential introduced in Sect. 2.6. It will also be heavily used in Chap. 3.

Two convenient references for Fréchet spaces and Fréchet manifolds are Hamilton [20] and Kriegl and Michor [42].

This is a good time to introduce some of our notational conventions. Spaces of mappings will usually have names beginning with capital letters, such as the diffeomorphism group $\mathrm{Diff}(M)$ or the space of embeddings $\mathrm{Emb}(V, M)$ of a submanifold of M. The same name beginning with a small letter, as in $\mathrm{diff}(M)$ or $\mathrm{emb}(V, M)$, will indicate the path component of the identity or inclusion map. We also use I to denote the standard unit interval [0, 1].

2.1 Fréchet Manifolds and the C^∞-Topology

For now, let M be a manifold with empty boundary. Throughout our work, we will use the C^∞-topology on $\mathrm{Diff}(M)$. For this topology, $\mathrm{Diff}(M)$ is a Fréchet manifold, locally diffeomorphic to the Fréchet space $\mathscr{Y}(M, TM)$ of smooth vector fields on M. In fact, the space $C^\infty(M, N)$ of smooth maps from M to N is a Fréchet manifold, metrizable when M is compact (see for example Theorem 42.1 and Proposition 42.3 of [42]), and $\mathrm{Diff}(M)$ is an open subset of $C^\infty(M, M)$ (Theorem 43.1 of [42]).

When M is compact, or more generally when one is working with maps and sections supported on a fixed compact subset of M, $\mathscr{Y}(M, TM)$ is a separable Fréchet space. By Theorem II.7.3 of [4], originally announced in [28], manifolds modeled on a separable Fréchet space Y are homeomorphic whenever they have the same homotopy type. Theorem IX.7.1 of [4] (originally Theorem 4 of [27])

S. Hong et al., *Diffeomorphisms of Elliptic 3-Manifolds*, Lecture Notes in Mathematics 2055, DOI 10.1007/978-3-642-31564-0_2,

shows that Diff(M) admits an open embedding into Y. Theorems II.6.2 and II.6.3 of [4] then show that Diff(M) has the homotopy type of a CW-complex. (As far as we know, this fact is due originally to Palais [52]; he showed that many infinite-dimensional manifolds are dominated by CW-complexes, but a space dominated by a CW-complex is homotopy equivalent to some CW-complex [44, Theorem IV.3.8].)

Fréchet manifolds are not the daily fare of low-dimensional topologists, and it is reasonable to ask what benefits their use brings to our work. Many of our results about the homotopy types of spaces of diffeomorphisms of three-manifolds are proven using algebraic methods which yield weak homotopy equivalences between spaces of diffeomorphisms and other spaces. To upgrade these weak homotopy equivalences to homotopy equivalences, one needs the fact that the spaces have the homotopy types of CW-complexes. In the fiber-preserving context, at least, we consider it far from adequate simply to assert that these are easy extensions of the case of diffeomorphism groups. Our use of Fréchet spaces provides a rather straightforward and unified method to prove that a wide variety of spaces of diffeomorphisms, embeddings, and images of embeddings have the homotopy types of CW-complexes.

Significantly, the upgrade of results from weak homotopy equivalence to homotopy equivalence extends very much further for Fréchet manifolds. As detailed earlier in this section, homotopy equivalent infinite-dimensional manifolds modeled on separable Fréchet spaces are actually homeomorphic. As discussed in Sect. 1.2, this allows us to find the actual homeomorphism type of many noteworthy spaces of diffeomorphisms.

Finally, the use of Fréchet spaces is very convenient, if perhaps not essential, in our study of the space of fibered structures on a singular fibered space (in particular, a Seifert-fibered three-manifold) in the final three sections of Chap. 3. For a singular fibering $p\colon \Sigma \to \mathcal{O}$, the space of fibered structures isomorphic to the given one is defined to be $\mathrm{Diff}(\Sigma)/\mathrm{Diff}_f(\Sigma)$. A key technical result, Theorem 3.12, is that the natural map $\mathrm{Diff}(\Sigma) \to \mathrm{Diff}(\Sigma)/\mathrm{Diff}_f(\Sigma)$ is a fibration. Its proof requires a measurement of when a fibered structure is close to being vertical. Such a measure is obtained in the proof of Theorem 3.12 by embedding $\mathrm{Diff}(\Sigma)/\mathrm{Diff}_f(\Sigma)$ into a certain Fréchet space of sections, giving a Fréchet manifold structure on $\mathrm{Diff}(\Sigma)/\mathrm{Diff}_f(\Sigma)$. The fibration in Theorem 3.12 is needed in Sect. 3.9 even to show that $\mathrm{Diff}(\Sigma)/\mathrm{Diff}_f(\Sigma)$ is weakly contractible. As before, then, the Fréchet structure upgrades this to actual contractibility.

2.2 Metrics Which are Products Near the Boundary

We are going to work extensively with manifolds with boundary, and will need special Riemannian metrics on them, which we develop in this section.

Recall that a Riemannian metric is called *complete* if every Cauchy sequence converges. For a complete Riemannian metric on M, a geodesic can be extended indefinitely unless it reaches a point in the boundary of M, where it may continue or it may fail to be extendible because it "runs out of the manifold."

Definition 2.1. A Riemannian metric on M is said to be a *product near the boundary* if there is a collar neighborhood $\partial M \times \mathrm{I}$ of the boundary on which the metric is the product of a complete metric on ∂M and the standard metric on I.

Note that when the metric is a product near the boundary, the exponential of any vector tangent to ∂M is a point in ∂M.

Given any collar $\partial M \times [0,2]$, it is easy to obtain a metric that is a product near the boundary of M. On $\partial M \times [0,2)$, fix a Riemannian metric that is the product of a metric on ∂M and the usual metric on $[0,2)$. Obtain the metric on M from this metric and any metric defined on all of M by using a partition of unity subordinate to the open cover $\{\partial M \times [0,2), M - \partial M \times \mathrm{I}\}$.

By a submanifold V of M, we mean a smooth submanifold. When M has boundary and $\dim(V) < \dim(M)$, we always require that V be properly embedded in the sense that $V \cap \partial M = \partial V$, and that every inward pointing tangent vector to V at a point in ∂V be also inward pointing in M.

We will often work with codimension-0 submanifolds of bounded manifolds. In that case, the submanifold is a *manifold with corners,* that is, locally diffeomorphic to a product of half-lines and lines. In fact, all of our work should extend straightforwardly into the full context of manifolds with corners, but for simplicity we restrict to the cases we will need. When V has codimension 0, we require that the frontier of V be a codimension-1 submanifold of M as above.

Definition 2.2. Suppose that the Riemannian metric on M is a product near the boundary, with respect to the collar $\partial M \times \mathrm{I}$. A submanifold V of M is said to *meet the collar $\partial M \times \mathrm{I}$ in I-fibers* when $V \cap \partial M \times \mathrm{I}$ is a union of I-fibers of $\partial M \times \mathrm{I}$.

Note that when V meets the collar of M in I-fibers, the normal space to V at any point (x,t) in $\partial M \times \mathrm{I}$ is contained in the subspace in $T_x M$ tangent to $\partial M \times \{t\}$. Consequently, if one exponentiates the $(<\epsilon)$-length vectors in the normal bundle to obtain a tubular neighborhood of V, then the fiber at (x,t) is contained in $\partial M \times \{t\}$.

Given a submanifold V, one may obtain a complete metric on M that is a product near ∂M and such that V meets the collar $\partial M \times \mathrm{I}$ in I-fibers as follows. First, obtain a collar of ∂M that V meets in I-fibers, by constructing an inward-pointing vector field on a neighborhood of ∂M which is tangent to V, using the integral curves associated to the vector field to produce the collar, then carrying out the previous construction to obtain a metric that is a product near the boundary for this collar. It is complete on the collar. To make it complete on all of M, define $f\colon M - \partial M \to (0,\infty)$ by putting $f(x)$ equal to the supremum of the values of r such that Exp is defined on all vectors in $T_x(M)$ of length less than r. Let $g\colon M - \partial M \to (0,\infty)$ be a smooth map that is an ϵ-approximation to $1/f$, and let $\phi\colon M \to [0,1]$ be a smooth map which is equal to 0 on $\partial M \times \mathrm{I}$ and is 1 on $M - \partial M \times [0,2)$. Give $M \times [0,\infty)$ the product metric, and define a smooth embedding $i\colon M \to M \times [0,\infty)$ by $i(x) = (x, \phi(x)g(x))$ if $x \notin \partial M$ and $i(x) = (x,0)$ if $x \in \partial M$. The restricted metric on $i(M)$ agrees with the product metric on $\partial M \times \mathrm{I}$ and is complete.

We will always assume that Riemannian metrics have been chosen to be complete.

2.3 Manifolds with Boundary

In this section, we will extend the results of Sect. 2.1 to the bounded case. We always assume that M has a Riemannian metric which is a product near the boundary for some collar $\partial M \times \mathrm{I}$.

Definition 2.3. Let V be a submanifold of M. By $\mathscr{Y}(V, TM)$ we denote the Fréchet space of all sections from V to the restriction of the tangent bundle of M to V. The zero section of $\mathscr{Y}(V, TM)$ is denoted by Z. For $L \subseteq M$, we denote by $\mathscr{Y}^L(V, TM)$ the subspace of $\mathscr{Y}(V, TM)$ consisting of the sections which equal Z on $V - L$.

The following extension lemma will be useful.

Lemma 2.1. *Form a manifold N from M and $\partial M \times (-\infty, 0]$ by identifying ∂M with $\partial M \times \{0\}$, and extending the metric on M using the product of the complete metric on ∂M and the standard metric on $(-\infty, 0]$.*

(i) *There is a continuous linear extension $E\colon \mathrm{C}^\infty(M, \mathbb{R}) \to \mathrm{C}^\infty(N, \mathbb{R})$ for which the image is contained in the subspace of functions that vanish on $\partial M \times (-\infty, -1]$.*
(ii) *There is a continuous linear extension $E\colon \mathscr{Y}(M, TM) \to \mathscr{Y}(N, TN)$ for which the image is contained in the subspace of sections that vanish on $\partial M \times (-\infty, -1]$.*

Proof. Part (i) is basically what is established in the proof of Corollary II.1.3.7 of [20]. It was also proven by essentially the same method, using series in place of integration and working only on a half-space in $\mathbb{R}^n$, by Seeley [61]. The extensions are first performed in local coordinates $\mathbb{R}^{n-1} \times \mathbb{R}$, where the value of the extension $Ef(x, t)$ for $t < 0$ is given by an integral on the ray $\{x\} \times [0, \infty)$. Fixing a collection of charts and a partition of unity, these local extensions are pieced together to give Ef. Multiplying by a smooth function which is 1 on a neighborhood of M and vanishes on $\partial M \times (-\infty, -1]$, we may achieve the final property in (i). Part (ii) follows from (i) since locally a vector field is just a collection of n real-valued functions. □

We are grateful to Tatsuhiko Yagasaki for bringing the reference [61] to our attention.

Our proof that $\mathrm{Diff}(M)$ is a Fréchet manifold will use the *tame exponential* TExp. Let X be a vector field on M such that for every $x \in M$, $\mathrm{Exp}(X(x))$ is defined. Then $\mathrm{TExp}(X)$ is defined to be the map from M to M that takes each x to $\mathrm{Exp}(X(x))$. For a complete manifold M without boundary, the tame exponential defines local charts on $\mathrm{C}^\infty(M, M)$ (and more generally on $\mathrm{C}^\infty(M, N)$ if instead of vector fields on M one uses sections of a pullback of TN to a bundle over M), see for example Theorem 42.1 of [42].

Definition 2.4. Let V be a submanifold of M, and as always assume that the metric on M is a product near the boundary and V meets $\partial M \times \mathrm{I}$ in I-fibers. By $\mathscr{X}(V, TM)$ we denote the Fréchet subspace of $\mathscr{Y}(V, TM)$ consisting of those sections which are tangent to ∂M at all points of $V \cap \partial M$. For $L \subseteq M$, we denote by $\mathscr{X}^L(V, TM)$ the subspace of sections that equal Z on $V - L$.

We remark that $\mathscr{X}(M, TM)$ is the tangent space at 1_M of the infinite-dimensional Lie group Diff(M), and the exponential map in that context takes a vector field on M to the map at time 1 of the flow on M associated to the vector field. The resulting exponential map from $\mathscr{X}(M, TM)$ to Diff(M) is not locally surjective near Z and 1_M, even for $M = S^1$ (see for example Sect. 5.5.2 of [20]). We will always use the tame exponential, which as noted above is a local homeomorphism (in fact, a local diffeomorphism, for appropriate structures on these spaces as infinite-dimensional manifolds).

We can now give the Fréchet structure on Diff(M). By $\mathrm{C}^\infty((M, \partial M), (M, \partial M))$ we will denote the space of smooth maps from M to M that take ∂M to ∂M, with the C^∞-topology.

Theorem 2.1. *The space* $\mathrm{C}^\infty((M, \partial M), (M, \partial M))$ *is a Fréchet manifold locally modeled on* $\mathscr{X}(M, TM)$*, and* Diff(M) *is an open subset of* $\mathrm{C}^\infty((M, \partial M), (M, \partial M))$.

Proof. It suffices to find a local chart for $\mathrm{C}^\infty((M, \partial M), (M, \partial M))$ at the identity 1_M that has image in Diff(M). Form a manifold N from M as in Lemma 2.1, and let $E \colon \mathscr{Y}(M, TM) \to \mathscr{Y}(N, TN)$ be a continuous linear extension as in part (ii) of Lemma 2.1. Since N is complete, the tame exponential TExp: $\mathscr{Y}(N, TN) \to \mathrm{C}^\infty(N, N)$ is defined. From Theorem 43.1 of [42], Diff(N) is an open subset of $\mathrm{C}^\infty(N, N)$. Let $U \subset \mathscr{Y}(N, TN)$ be an open neighborhood of Z which TExp carries homeomorphically to an open neighborhood of 1_N in Diff(N). Since vector fields in $\mathscr{X}(M, TM)$ are tangent to the boundary, TExp carries $U \cap E(\mathscr{X}(M, TM))$ to diffeomorphisms of N taking ∂M to ∂M. Therefore TExp carries the open neighborhood $E^{-1}(U \cap E(\mathscr{X}(M, TM)))$ of Z into Diff(M). □

As in the case of manifolds without boundary, we can now conclude that Diff(M) has the homotopy type of a CW-complex.

When M is compact, Diff(M) is separable, and moreover Diff(M) is locally convex. Explicitly, our local charts defined using the tame exponential show that for any $f \in$ Diff(M), there is a neighborhood U of f such that for every $g \in U$, the homotopy that moves points along the shortest geodesic from each $g(x)$ to $f(x)$ is an isotopy from g to f.

For a closed subset $X \subset M$, we denote by Diff(M rel X) the subgroup of Diff(M) consisting of the elements which take X to X and restrict to the identity map on X. Adapting the previous arguments shows that Diff(M rel X) is modeled on the closed Fréchet subspace of $\mathscr{X}(M, TM)$ consisting of sections that vanish on X.

2.4 Spaces of Embeddings

When we work with embeddings, we always start with a fixed submanifold V of the ambient manifold M. The inclusion map then furnishes a natural basepoint of the space of embeddings. In addition, this will allow a simple definition of the space of images of V in M, given in Definition 3.5 below.

Definition 2.5. Let V be a submanifold of M. When M has boundary and $\dim(V) < \dim(M)$, we always require that $V \cap \partial M = \partial V$, and select our Riemannian metric on M to be a product near the boundary for which V meets the collar $\partial M \times \mathrm{I}$ in I-fibers. Similarly, when V is codimension-0, the frontier of V is a codimension-1 submanifold of M assumed to meet $\partial M \times \mathrm{I}$ in I-fibers. Denote by $\mathrm{Emb}(V, M)$ the space of all smooth embeddings j of V into M such that

(i) $j^{-1}(\partial M) = V \cap \partial M$
(ii) j extends to a diffeomorphism from M to M.

Note that condition (ii) implies that j carries every inward-pointing tangent vector of $V \cap \partial M$ to an inward-pointing tangent vector of M. It also implies that the natural map $\mathrm{Diff}(M) \to \mathrm{Emb}(V, M)$ that sends each diffeomorphism to its restriction to V is surjective.

With the C^∞-topology, $\mathrm{Emb}(V, M)$ is a Fréchet manifold locally modeled on $\mathscr{X}(V, TM)$. For the closed case, this is proven in Theorem 44.1 of [42], and adaptations like those in Sect. 2.3 allow its extension in the bounded and codimension-0 contexts (note that Lemma 2.1 provides a continuous linear extension from $\mathscr{X}(V, M)$ to $\mathscr{X}(V \cup (-\infty, 0], M \cup (-\infty, 0]))$. As in the case of $\mathrm{Diff}(M)$, this Fréchet manifold structure shows that $\mathrm{Emb}(V, M)$ has the homotopy type of a CW-complex.

2.5 Bundles and Fiber-Preserving Diffeomorphisms

Let $p\colon E \to B$ be a locally trivial smooth map of manifolds, with compact fiber. When B and the fiber have nonempty boundary, E should be regarded as a manifold with corners at the boundary points of the fibers in $p^{-1}(\partial B)$. The horizontal boundary $\partial_h E$ is defined to be $\cup_{x \in B} \partial(p^{-1}(x))$, and the vertical boundary $\partial_v E$ to be $p^{-1}(\partial B)$.

Definition 2.6. The space of *fiber-preserving diffeomorphisms* $\mathrm{Diff}_f(E)$ is the subspace of $\mathrm{Diff}(E)$ consisting of the diffeomorphisms that take each fiber of E to a fiber. The *vertical diffeomorphisms* $\mathrm{Diff}_v(E)$ are the elements of $\mathrm{Diff}_f(E)$ that take each fiber to itself.

Fibered submanifolds also play an important role.

Definition 2.7. A submanifold W of E is called *fibered* or *vertical* if it is a union of fibers. For a fibered submanifold W of E, define $\partial_h W$ to be $W \cap \partial_h E$ and $\partial_v W$ to be $W \cap \partial_v E$. The space of *fiber-preserving embeddings* $\mathrm{Emb}_f(W, E)$ is the subspace of $\mathrm{Emb}(W, E)$ consisting of embeddings that take each fiber of W to a fiber of E, and the space of *vertical embeddings* $\mathrm{Emb}_v(W, E)$ is the subspace of $\mathrm{Emb}_f(W, E)$ consisting of embeddings taking each fiber to itself.

At each point $x \in E$, let $V_x(E)$ denote the *vertical subspace* of $T_x(E)$ consisting of vectors tangent to the fiber of p. When E has a Riemannian metric, the orthogonal complement $H_x(E)$ of $V_x(E)$ in $T_x(E)$ is called the *horizontal subspace.* We call the elements of $V_x(E)$ and $H_x(E)$ *vertical* and *horizontal* respectively. Clearly $V_x(E)$ is the kernel of $p_*\colon T_x(E) \to T_{p(x)}(B)$, while $p_*|_{H_x(E)}\colon H_x(E) \to T_{p(x)}(B)$ is an isomorphism. Each vector $\omega \in T_x(E)$ has an orthogonal decomposition $\omega = \omega_v + \omega_h$ into its vertical and horizontal parts.

A path α in E is called *horizontal* if $\alpha'(t) \in H_{\alpha(t)}(E)$ for all t in the domain of α. Let $\gamma\colon [a, b] \to B$ be a path such that $\gamma'(t)$ never vanishes, and let $x \in E$ with $p(x) = \gamma(a)$. A horizontal path $\widetilde{\gamma}\colon [a, b] \to E$ such that $\widetilde{\gamma}(a) = x$ and $p\widetilde{\gamma} = \gamma$ is called a *horizontal lift* of γ starting at x.

To ensure that horizontal lifts exist, we will need a special metric on E.

Definition 2.8. A Riemannian metric on E is said to be a *product near* $\partial_h E$ when

(i) There is a collar neighborhood $\partial_h E \times \mathrm{I}$ of the horizontal boundary on which the metric is the product of a complete metric on $\partial_h E$ and the standard metric on I.
(ii) For this collar $\partial_h E \times \mathrm{I}$, each $\{x\} \times \mathrm{I}$ lies in some fiber of p.

Such metrics can be constructed using a partition of unity as follows. Using the local product structure, at each point x in $\partial_h E$ select a vector field defined on a neighborhood of x that

(a) Points into the fiber at points of $\partial_h E$
(b) Is tangent to the fibers wherever it is defined.

By (b), the vector field must be tangent to $\partial_v E$ at points in $\partial_v E$. Since scalar multiples and linear combinations of vectors satisfying these two conditions also satisfy them, we may piece these local fields together using a partition of unity to construct a vector field, nonvanishing on a neighborhood of $\partial_h E$, that satisfies (a) and (b). Using the integral curves associated to this vector field we obtain a smooth collar neighborhood $\partial_h E \times [0, 2]$ of $\partial_h E$ such that each $[0, 2]$-fiber lies in a fiber of p. On $\partial_h E \times [0, 2)$, fix a Riemannian metric that is the product of a metric on $\partial_h E$ and the usual metric on $[0, 2)$. Form a metric on E from this metric and any metric on all of E using a partition of unity subordinate to the open cover $\{\partial_h E \times [0, 2), E - \partial_h E \times \mathrm{I}\}$.

When the metric is a product near $\partial_h E$ such that the I-fibers of $\partial_h E \times \mathrm{I}$ are vertical, the horizontal subspace H_x is tangent to $\partial_h E \times \{t\}$ whenever $x \in \partial_h E \times \{t\}$. For H_x is orthogonal to the fiber $p^{-1}(p(x))$, and since the I-fiber of $\partial_h E \times \mathrm{I}$ that contains x lies in $p^{-1}(p(x))$, H_x is orthogonal to that I-fiber as well. Since $\partial_h E \times \{t\}$ meets the I-fiber orthogonally, with codimension 1, H_x is tangent to $\partial_h E \times \{t\}$.

Since the horizontal subspaces are tangent to the $\partial_h E \times \{t\}$, a horizontal lift starting in some $\partial_h E \times \{t\}$ will continue in $\partial_h E \times \{t\}$. Provided that the fiber is compact, as we are assuming, the existence of horizontal lifts is assured.

2.6 Aligned Vector Fields and the Aligned Exponential

For working with fiber-preserving diffeomorphisms and embeddings, we will use a variant of the exponential map, called the *aligned exponential.* It behaves nicely with respect to aligned vector fields, which are the vector fields that project to a well-defined vector field in the base manifold. Precisely, we have:

Definition 2.9. A vector field $X\colon E \to TE$ is called *aligned* if $p(x) = p(y)$ implies that $p_*(X(x)) = p_*(X(y))$ (these are often called *projectable* in the literature). This happens precisely when there exist a vector field X_B on B and a vertical vector field X_V on E so that for all $x \in E$,

$$X(x) = (p_*|_{H_x})^{-1}(X_B(p(x))) + X_V(x)\,.$$

In particular, any vertical vector field is aligned. When X is aligned, the projected vector field p_*X is well-defined.

The idea of the aligned exponential Exp_a is that it behaves as would the regular exponential if the metric on E were locally the product of a metric on F and a metric on B. The key property of Exp_a is that if X is an aligned vector field on E, and $\mathrm{Exp}_a(X(x))$ is defined for all x, then the map of E defined by sending x to $\mathrm{Exp}_a(X(x))$ will be fiber-preserving.

Definition 2.10. Let $\pi\colon TE \to E$ denote the tangent bundle of E. Assume that the metric on E is a product near $\partial_h E$ such that the I-fibers of $\partial_h E \times \mathrm{I}$ are vertical. Each fiber F of E inherits a Riemannian metric from that of E, and has an exponential map Exp_F which (where defined) carries vectors tangent to F to points of F. The path $\mathrm{Exp}_F(t\omega)$ is not generally a geodesic in E. The *vertical exponential* Exp_v is defined by $\mathrm{Exp}_v(\omega) = \mathrm{Exp}_F(\omega)$, where ω is a vertical vector and F is the fiber containing $\pi(\omega)$. The *aligned exponential map* Exp_a is defined as follows. Consider a tangent vector $\omega \in T_x(E)$ such that for the vector $p_*(\omega) \in T_{p(x)}(B)$, $\mathrm{Exp}(p_*(\omega))$ is defined. A geodesic segment $\gamma_{p_*(\omega)}$ starting at $p(\pi(\omega))$ is defined by $\gamma_{p_*(\omega)}(t) = \mathrm{Exp}(tp_*(\omega))$, $0 \leq t \leq 1$. Define $\mathrm{Exp}_a(\omega)$ to be the endpoint of the unique horizontal lift of $\gamma_{p_*(\omega)}$ starting at $\mathrm{Exp}_v(\omega_v)$.

Note that $\mathrm{Exp}_a(\omega)$ exists if and only if both $\mathrm{Exp}_v(\omega_v)$ and $\mathrm{Exp}(p_*(\omega))$ exist. Clearly, when $\mathrm{Exp}_a(\omega)$ is defined, it lies in the fiber containing the endpoint of a lift of $\gamma_{p_*(\omega)}$, and therefore $p(\mathrm{Exp}_a(\omega)) = \mathrm{Exp}(p_*(\omega))$. This immediately implies that if X is an aligned vector field on E such that $\mathrm{Exp}_a(X(x))$ is defined for all $x \in E$, then the map defined by sending x to $\mathrm{Exp}_a(X(x))$ takes fibers to fibers, and in particular if X is vertical, it takes each fiber to itself.

Definition 2.11. Let W be a vertical submanifold of E. By $\mathscr{A}(W, TE)$ we denote the Fréchet space of sections X from W to $TE|_W$ such that

(1) X is aligned, that is, if $p(w_1) = p(w_2)$ then $p_*(X(w_1)) = p_*(X(w_2))$.
(2) If $x \in \partial_h W$, then $X(x)$ is tangent to $\partial_h E$, and if $x \in \partial_v W$, then $X(x)$ is tangent to $\partial_v E$.

The elements of $\mathscr{A}(W, TE)$ such that $p_* X(x) = Z(p(x))$ for all $x \in W$ are denoted by $\mathscr{V}(W, TE)$.

The vector space structure on $\mathscr{A}(W, TE)$ is defined using the vector space structures of the fibers of TE and TB. Given $v, w \in \mathscr{A}(W, TE)$, we decompose them into their vertical and horizontal parts. The vertical parts are added by the usual addition in TE. The horizontal parts are added by pushing down to TB, adding there, and taking horizontal lifts.

Since horizontal lifts of geodesics in B exist, $\mathrm{Exp}_a(\omega)$ is defined whenever $\mathrm{Exp}_v(\omega)$ and $\mathrm{Exp}(p_*(\omega))$ are defined. In particular, the tame aligned exponential TExp_a carries a neighborhood of Z in $\mathscr{A}(W, TE)$ into $C^\infty_f(W, E)$. Choosing the neighborhood small enough to ensure that $\mathrm{TExp}_a(X) \in \mathrm{Emb}_f(W, E)$ provides local charts on $\mathrm{Emb}_f(W, E)$, that carry the vertical fields into $\mathrm{Emb}_v(W, E)$. Thus we have:

Theorem 2.2. *The spaces* $\mathrm{Diff}_f(E)$, $\mathrm{Diff}_v(E)$, $\mathrm{Emb}_f(W, E)$, *and* $\mathrm{Emb}_v(W, E)$ *are infinite-dimensional manifolds modeled on Fréchet spaces of aligned vector fields.*

Chapter 3
The Method of Cerf and Palais

In [51], Palais proved a very useful result relating diffeomorphisms and embeddings. For closed M, it says that if $W \subseteq V$ are submanifolds of M, then the mappings $\mathrm{Diff}(M) \to \mathrm{Emb}(V, M)$ and $\mathrm{Emb}(V, M) \to \mathrm{Emb}(W, M)$ obtained by restricting diffeomorphisms and embeddings are locally trivial, and hence are Serre fibrations. The same results, with variants for manifolds with boundary and more complicated additional boundary structure, were proven by Cerf in [10]. Among various applications of these results, the Isotopy Extension Theorem follows by lifting a path in $\mathrm{Emb}(V, M)$ starting at the inclusion map of V to a path in $\mathrm{Diff}(M)$ starting at 1_M. Moreover, parameterized versions of isotopy extension follow just as easily from the homotopy lifting property for $\mathrm{Diff}(M) \to \mathrm{Emb}(V, M)$ (see Corollary 3.3).

In this chapter, we will extend the theorem of Palais in various ways. Many of our results concern fiber-preserving maps. For example, in Sect. 3.3 we will prove the

Projection Theorem (Theorem 3.4) *Let E be a bundle over a compact manifold B. Then* $\mathrm{Diff}_f(E) \to \mathrm{Diff}(B)$ *is locally trivial.*

This should be considered a folk theorem. Below we will discuss some of its antecedents.

The homotopy extension property for the projection fibration $\mathrm{Diff}_f(E) \to \mathrm{Diff}(B)$ translates directly into the following.

Parameterized Isotopy Lifting Theorem (Corollary 3.3) *Suppose that $p\colon E \to B$ is a fibering of compact manifolds, and suppose that for each t in a path-connected parameter space P, there is an isotopy $g_{t,s}$ such that $g_{t,0}$ lifts to a diffeomorphism $G_{t,0}$ of E. Assume that sending $(t, s) \to g_{t,s}$ defines a continuous function from $P \times [0, 1]$ to* $\mathrm{Diff}(B)$ *and sending t to $G_{t,0}$ defines a continuous function from P to* $\mathrm{Diff}_f(E)$. *Then the family $G_{t,0}$ extends to a continuous family on $P \times \mathrm{I}$ such that for each (t, s), $G_{t,s}$ is a fiber-preserving diffeomorphism inducing $g_{t,s}$ on B.*

For fiber-preserving and vertical embeddings of vertical submanifolds, we have a more direct analogue of Palais' results.

S. Hong et al., *Diffeomorphisms of Elliptic 3-Manifolds*, Lecture Notes in Mathematics 2055, DOI 10.1007/978-3-642-31564-0_3,

Restriction Theorem (Corollaries 3.4 and 3.5) *Let* V *and* W *be vertical submanifolds of* E *with* $W \subseteq V$*, each of which is either properly embedded or codimension-zero. Then the restrictions* $\mathrm{Diff}_f(M) \to \mathrm{Emb}_f(V, M)$, $\mathrm{Diff}_v(M) \to \mathrm{Emb}_v(V, M)$, $\mathrm{Emb}_f(V, E) \to \mathrm{Emb}_f(W, E)$ *and* $\mathrm{Emb}_v(V, E) \to \mathrm{Emb}_v(W, E)$ *are locally trivial.*

As shown in Theorem 3.6, the Projection and Restriction Theorems can be combined into a single commutative square, called the *projection-restriction square,* in which all four maps are locally trivial:

$$\begin{array}{ccc} \mathrm{Diff}_f(E) & \longrightarrow & \mathrm{Emb}_f(W, E) \\ \downarrow & & \downarrow \\ \mathrm{Diff}(B) & \longrightarrow & \mathrm{Emb}(p(W), B)\,. \end{array}$$

In three-dimensional topology, a key role is played by manifolds admitting a more general kind of fibered structure, called a Seifert fibering. Some general references for Seifert-fibered three-manifolds are [26, 37, 38, 49, 50, 60, 62, 69, 70]. In Sect. 3.6, we prove the analogues of the results discussed above for most Seifert fiberings $p\colon \Sigma \to \mathcal{O}$. Actually, we work in a somewhat more general context, called *singular fiberings*, which resemble Seifert fiberings but for which none of the usual structure of the fiber as a homogeneous space is required.

In the late 1970s fibration results akin to our Projection Theorem for the singular fibered case were proven by Neumann and Raymond [48]. They were interested in the case when Σ admits an action of the k-torus T^k and $\Sigma \to \mathcal{O}$ is the quotient map to the orbit space of the action. They proved that the space of (weakly) T^k-equivariant homeomorphisms of Σ fibers over the space of homeomorphisms of $\mathcal{O}$ that respect the orbit types associated to the points of $\mathcal{O}$. A detailed proof of this result when the dimension of Σ is $k + 2$ appears in the dissertation of Park [53]. Park also proved analogous results for space of weakly G-equivariant maps for principal G-bundles and for Seifert fiberings of arbitrary dimension [53, 54]. These results do not directly overlap ours since we always consider the full group of fiber-preserving diffeomorphisms without any restriction to G-equivariant maps (indeed, no assumption of a G-action is even present).

The results of this chapter will be used heavily in the later chapters. In this chapter, we give one main application. For a Seifert-fibered manifold Σ, $\mathrm{Diff}(\Sigma)$ acts on the set of Seifert fiberings, and the stabilizer of the given fibering is $\mathrm{Diff}_f(\Sigma)$, thus the space of cosets $\mathrm{Diff}(\Sigma)/\,\mathrm{Diff}_f(\Sigma)$ can be regarded as the *space of Seifert fiberings* of Σ equivalent to the given one. We prove in Sect. 3.9 that for a Seifert-fibered Haken three-manifold, each component of the space of Seifert fiberings is contractible (apart from a small list of well-known exceptions, the space of Seifert fiberings is connected). This too should be considered a folk result; it appears to be widely believed, and regarded to be a direct consequence of the work of Hatcher and Ivanov on the diffeomorphism groups of Haken manifolds. We have found, however, that a real proof requires more than a little effort.

Our results will be proven by adapting the Palais method of [51], using the aligned exponential defined in Sect. 2.6. In Sect. 3.1, we reprove the main result of [51] for manifolds which may have boundary. This duplicates [10] (in fact, the boundary control there is more refined than ours), but is included to furnish lemmas as well as to exhibit a prototype for the approach we use to deal with the bounded case in our later settings. In Sect. 3.5, we give the analogues of the results of Palais and Cerf for smooth orbifolds, which for us are quotients $\widetilde{\mathscr{O}}/H$ where $\widetilde{\mathscr{O}}$ is a manifold and H is a group acting smoothly and properly discontinuously on $\widetilde{\mathscr{O}}$. Besides being of independent interest, these analogues are needed for the case of singular fiberings.

Throughout this chapter, all Riemannian metrics are assumed to be products near the boundary, or near the horizontal boundary for total spaces of bundles, such that any submanifolds under consideration meet the collars in I-fibers. Let V be a submanifold of M. As in Definition 2.4, the notation $\mathscr{X}(V, TM)$ means the Fréchet space of sections from V to the restriction of the tangent bundle of M to V that are tangent to ∂M at all points of $V \cap \partial M$. We also utilize various kinds of control, as indicated in the following definitions.

Definition 3.1. The notations $\mathrm{Diff}(M \text{ rel } X)$ and $\mathrm{Diff}^{M-X}(M)$ mean the space of diffeomorphisms which restrict to the identity map on each point of the subset X of M. These notations may be combined, for example $\mathrm{Diff}^{L}(M \text{ rel } X)$ is the space of diffeomorphisms that are the identity on $X \cup (M - L)$.

Definition 3.2. For $X \subseteq M$ we say that $K \subseteq M$ is a neighborhood of X when X is contained in the topological interior of K. If K is a neighborhood of a submanifold V of M, then $\mathrm{Emb}^{K}(V, M)$ means the elements j in $\mathrm{Emb}(V, M)$ such that K is a neighborhood of $j(V)$. Suppose that S is a closed neighborhood in ∂M of $V \cap S$. Note that this implies that $S \cap \partial V$ is a union of components of $V \cap \partial M$. We denote by $\mathrm{Emb}(V, M \text{ rel } S)$ the elements j that equal the inclusion on $V \cap S$ and carry $V \cap (\partial M - S)$ into $\partial M - S$. For a neighborhood K of V, the superscript notation of Definition 3.1 may be used, as in $\mathrm{Emb}^{K}(V, M \text{ rel } S)$.

Definition 3.3. Recall from Definition 2.4 that for $L \subseteq M$, $\mathscr{X}^{L}(V, TM)$ means the elements of $\mathscr{X}(V, TM)$ that equal the zero section Z on $V - L$. We extend this to the aligned and vertical sections (see Definition 2.11), so that if $L \subset E$ then $\mathscr{A}^{L}(W, TE)$ and $\mathscr{V}^{L}(W, TE)$ and $\mathscr{V}^{L}(W, TE)$ have the corresponding meanings.

3.1 The Palais–Cerf Restriction Theorem

We begin with a review of the method of Palais [51].

Definition 3.4. Let X be a G-space and $x_0 \in X$. A *local cross-section* (or *G local cross-section*) for X at x_0 is a map χ from a neighborhood U of x_0 into G such that $\chi(u)x_0 = u$ for all $u \in U$. By replacing $\chi(u)$ by $\chi(u)\chi(x_0)^{-1}$, one may always assume that $\chi(x_0) = 1_G$. If X admits a local cross-section at each point, it is said to admit local cross-sections.

Note that a local cross-section $\chi_0: U_0 \to G$ at a single point x_0 determines a local cross section $\chi: gU_0 \to G$ at any point gx_0 in the orbit of x_0, by the formula $\chi(u) = g\chi_0(g^{-1}u)g^{-1}$, since then $\chi(u)(gx_0) = g\chi_0(g^{-1}u)g^{-1}gx_0 = g\chi_0(g^{-1}u)x_0 = gg^{-1}u = u$. In particular, if G acts transitively on X, then a local cross section at any point provides local cross sections at all points.

From [51] we have

Proposition 3.1. *Let G be a topological group and X a G-space admitting local cross-sections. Then any equivariant map of a G-space into X is locally trivial.*

In fact, when $\pi: Y \to X$ is G-equivariant, the local coordinates on $\pi^{-1}(U)$ are just given by sending the point $(u, z) \in U \times \pi^{-1}(y_0)$ to $\chi(u) \cdot z$. Some additional properties of the bundles obtained in Proposition 3.1 are given in [51].

Example 3.1. For a closed subgroup H of a Lie group G, the projection $G \to G/H$ to the space of left cosets of H always has local G cross-sections, and hence is locally trivial. To check this, recall first that since G acts transitively on G/H, it is sufficient to find a local cross-section χ_0 at the coset $e\,H$, where e is the identity element of G. To construct χ_0, fix a Riemannian metric on G. The tangent space T_eH is a subspace of T_eG. Let W be a complementary subspace. Let V be an open neighborhood of 0 in T_eG such that Exp: $V \to U$ is a diffeomorphism onto an open neighborhood of e in G, and so that the submanifold Exp$(W \cap V)$ is transverse to the cosets uH for all $u \in U$. Defining $\chi_0(uH)$ to be Exp(w) for the unique element $w \in W \cap U$ such that Exp$(w)U = uH$ gives the local cross-section at e.

The following technical lemma will simplify some of our applications of Proposition 3.1.

Proposition 3.2. *Let M be a G-space and let V be a subspace of M, possibly equal to M. Let $I(V, M)$ be a space of embeddings of V into M, on which G acts by composition on the left. Suppose that for every $i \in I(V, M)$, the space of embeddings $I(i(V), M)$ has a local G cross-section at the inclusion map of $i(V)$ into M. Then $I(V, M)$ has local G cross-sections.*

Proof. Fix $i \in I(V, M)$, and denote by $j_{i(V)}$ the inclusion map of $i(V)$ into M. Define $Y: I(V, M) \to I(i(V), M)$ by $Y(j) = ji^{-1}$. For a local cross-section $\chi: U \to G$ at $j_{i(V)}$, define Y_1 to be the restriction of Y to $Y^{-1}(U)$, a neighborhood of i in $I(V, M)$. Then $\chi Y_1: Y^{-1}(U) \to G$ is a local cross-section for $I(V, M)$ at i. For if $j \in Y^{-1}(U)$, then $\chi(Y_1(j)) \circ i = \chi(Y_1(j)) \circ j_{i(V)} \circ i = Y_1(j) \circ i = j$. □

In our context, a typical procedure for finding a local cross-section using the Palais method is as follows. Suppose, for example, that one wants to find a local cross-section from a space of embeddings of a submanifold to a space of diffeomorphisms of the ambient manifold. First, take the "logarithm" of an embedding j, that is, find a section from the submanifold to the tangent bundle of M so that the exponential of the vector at each x is the image $j(x)$. Then obtain an "extension" of this section to a vector field on M. Finally, "exponentiate" the extended vector field to obtain the diffeomorphism of M that agrees with j on the submanifold. The extension process must be canonical enough so that sending the embedding to the resulting diffeomorphism is a local cross-section.

This three-step procedure depends in large part on three lemmas, or appropriate versions of them, called Lemmas d, c and b in [51]. As our first instance of them, we give the following versions for manifolds with boundary and submanifolds that may be of codimension 0.

Lemma 3.1 (Logarithm Lemma). *Assume that the metric on M is a product near ∂M, and let V be a compact submanifold of M that meets $\partial M \times \mathrm{I}$ in* I*-fibers. Then there are an open neighborhood U of the inclusion i_V in* $\mathrm{Emb}(V, M)$ *and a continuous map $X\colon U \to \mathscr{X}(V, TM)$ such that for all $j \in U$,* $\mathrm{Exp}(X(j)(x))$ *is defined for all $x \in V$ and* $\mathrm{Exp}(X(j)(x)) = j(x)$ *for all $x \in V$. Moreover, $X(i_V) = Z$.*

Proof. Choose ε small enough so that for all $x \in V$, Exp carries the ε-ball about 0 in $T_x(M)$ (that is, the portion of this ε-ball on which it is defined, which may be as small as a closed half-ball for $x \in \partial M$) diffeomorphically to a neighborhood W_x of x in M. Choose a neighborhood U of i_V in $\mathrm{Emb}(V, M)$ so that if $j \in U$ then $j(x) \in W_x$. For $j \in U$ define $X(j)(x)$ to be the unique vector in $T_x(M)$ of length less than ε for which $\mathrm{Exp}(X(j)(x))$ equals $j(x)$. Since the metric is a product near the boundary, $X(j)$ is in $\mathscr{X}(V, M)$, and the remark about i_V is clear. □

The Extension Lemma uses the notation from Definition 3.3.

Lemma 3.2 (Extension Lemma). *Assume that the metric on M is a product near ∂M, and let V be a compact submanifold of M that meets $\partial M \times \mathrm{I}$ in* I*-fibers. Let L be a neighborhood of V in M. Then there exists a continuous linear map $k\colon \mathscr{X}(V, TM) \to \mathscr{X}^L(M, TM)$ such that $k(X)(x) = X(x)$ for all x in V and all X in $\mathscr{X}(V, TM)$, and moreover if S is a closed neighborhood in ∂M of $S \cap \partial V$, and if $X(x) = Z(x)$ for all $x \in S \cap \partial V$, then $k(X)(x) = Z(x)$ for all $x \in S$.*

Proof. Suppose first that V has positive codimension. Let $\nu_\varepsilon(V)$ denote the subspace of the normal bundle of V consisting of vectors of length ε, and let $e\colon \nu_\varepsilon(V) \to M$ be the exponentiation map. For ε sufficiently small, e is a diffeomorphism onto a tubular neighborhood of V in M. Since the metric on M is a product near the boundary, and V meets $\partial M \times \mathrm{I}$ in I-fibers, the fibers of $\nu_\varepsilon(V)$ are carried into the submanifolds $\partial M \times \{t\}$ near the boundary.

Suppose that $v \in T_x(M)$ and that $\mathrm{Exp}(v)$ is defined. For all $u \in T_x(M)$ define $P(u, v)$ to be the vector that results from parallel translation of u along the path that sends t to $\mathrm{Exp}(tv)$, $0 \le t \le 1$. In particular, $P(u, Z(x)) = u$ for all u. Let $\alpha\colon M \to [0, 1]$ be a smooth function which is identically 1 on V and identically 0 on $M - e(\nu_{\varepsilon/2}(V))$. Define $k\colon \mathscr{X}(V, TM) \to \mathscr{X}^L(M, TM)$ by

$$k(X)(x) = \begin{cases} \alpha(x) P(X(\pi(e^{-1}(x))), e^{-1}(x)) & \text{for } x \in e(\nu_\varepsilon(V)) \\ Z(x) & \text{for } x \in M - e(\nu_{\varepsilon/2}(V)) \end{cases}$$

For $x \in V$, $e^{-1}(x) = Z(x)$ and $\alpha(x) = 1$, so $k(X)(x) = X(x)$. Similarly, $k(X)(x) = Z(x)$ for $x \in M - L$. For $x \in \partial M$, $\pi(e^{-1}(x))$ is also in ∂M, so $X(\pi(e^{-1}(x)))$ is tangent to the boundary. Since the metric is a product near the boundary, $P(X(\pi(e^{-1}(x))), e^{-1}(x))$ is also tangent to the boundary. Therefore $k(X) \in \mathscr{X}^L(M, TM)$.

Assume now that V has codimension zero, so that its frontier W is a properly embedded submanifold. Fix a tubular neighborhood $W \times (-\infty, \infty)$, contained in L, with $V \cap (W \times (-\infty, \infty)) = W \times [0, \infty)$. As in Lemma 2.1, there is a continuous linear extension map $E\colon \mathscr{X}(V, TM) \to \mathscr{Y}(V \cup (W \times (-\infty, \infty)), TM)$. Note that since M may have boundary, it is necessary to use the half-space version of reference [61] at points of $V \cap \partial M$. The extended vector fields are Z on $W \times [1, \infty)$, so extend using Z on $M - (V \cup (W \times (-\infty, \infty)))$. At points of ∂M, the component of each vector in the direction perpendicular to ∂M is 0, so since E is linear, the extended component is also 0 and therefore the extended vector field is also tangent to the boundary. This defines $k\colon \mathscr{X}(V, TM) \to \mathscr{X}^L(M, TM)$.

The final sentence of the proof holds provided that we choose the tubular neighborhoods small enough to have fibers contained in S at points in $V \cap \partial M$, or in $\partial M - S$ at points in $V \cap (\partial M - S)$. □

Lemma 3.3 (Exponentiation Lemma). *Assume that the metric on M is a product near the boundary, and let K be a compact subset of M. Then there exists a neighborhood U of Z in $\mathscr{X}^K(M, TM)$ such that* TExp(X) *is defined for all $X \in U$, and* TExp *carries U into* Diff^K M).

Proof. Form a manifold N from M and $\partial M \times (-\infty, 0]$ by identifying ∂M with $\partial M \times \{0\}$, and extending the metric on M using the product of the complete metric on ∂M and the standard metric on $(-\infty, 0]$. By Lemma 2.1, there is a continuous linear extension $E\colon \mathscr{Y}(M, TM) \to \mathscr{Y}(N, TN)$ for which the image is contained in the subspace of sections that vanish on $\partial M \times (-\infty, -1]$. Put $L = K \cup (K \cap \partial M) \times [-1, 0]$. As seen in the proof of Lemma 2.1, the extended vector fields may be chosen to lie in $\mathscr{Y}^L(N, TN)$. Since N is complete and open, there is a neighborhood W of Z in $\mathscr{Y}^L(N, TN)$ for which $\mathrm{Exp}(E(Y(x)))$ is defined for all $Y \in W$ and $x \in N$. That is, TExp: $W \to C^\infty(N, N)$ is defined.

Since Diff(N) is an open subset of $C^\infty(N, N)$, W may be chosen smaller, if necessary, to ensure that it is carried into Diff(N) by TExp. Diffeomorphisms obtained from extended vector fields of $\mathscr{X}(M, TM)$ carry ∂M to ∂M, so TExp carries the neighborhood $U = \mathscr{X}^K(M, TM) \cap E^{-1}(W)$ of Z in $\mathscr{X}^K(M, TM)$ into Diff$^K(M)$. □

We are now set up for the main results of this section. At this point the reader may wish to review Definitions 3.1 and 3.2.

Theorem 3.1. *Let V be a compact submanifold of M, and let S be a closed neighborhood in ∂M of $S \cap \partial V$. Let L be a compact neighborhood of V in M. Then* Emb$^L(V, M$ rel $S \cap \partial V)$ *admits local* Diff$^L(M$ rel $S)$ *cross-sections.*

Proof. By Proposition 3.2 it suffices to find a local cross-section at the inclusion map i_V. Using Lemmas 3.1 and 3.2, we obtain an open neighborhood W of i_V in $\mathrm{Emb}(V, M \text{ rel } S \cap \partial V)$ and continuous maps $X: W \to \mathscr{X}(V, TM)$ and $k: \mathscr{X}(V, TM) \to \mathscr{X}^L(M, TM)$. By Lemma 3.3, there is a neighborhood U of Z in $\mathscr{X}^L(M, TM)$ for which the map $F: U \to \mathrm{Diff}^L(M)$ sending Y to $\mathrm{TExp}(Y)$ is defined and continuous. Choosing a neighborhood U_1 of i_V contained in $W \cap (k \circ X)^{-1}(U)$, the function $F \circ k \circ X: U_1 \to \mathrm{Diff}^L(M \text{ rel } S)$ will be the desired cross-section.

To see that the image of this function lies in $\mathrm{Diff}^L(M \text{ rel } S)$, suppose that $j(x) = x$ for all $x \in V \cap S$. Then $X(j)(x) = Z(x)$ for $x \in V \cap S$. By the condition in Lemma 3.2, $k(X(j))(x) = Z(x)$ and consequently $(FkX(j))(x) = x$ for all $x \in S$. □

Using Proposition 3.1 we obtain immediate corollaries of Theorem 3.1:

Corollary 3.1. *Let V be a compact submanifold of M. Let $S \subseteq \partial M$ be a closed neighborhood in ∂M of $S \cap \partial V$, and L a neighborhood of V in M. Then the restriction* $\mathrm{Diff}^L(M \text{ rel } S) \to \mathrm{Emb}^L(V, M \text{ rel } S)$ *is locally trivial.*

Corollary 3.2. *Let V and W be compact submanifolds of M, with $W \subseteq V$. Let $S \subseteq \partial M$ a closed neighborhood in ∂M of $S \cap \partial V$, and L a neighborhood of V in M. Then the restriction* $\mathrm{Emb}^L(V, M \text{ rel } S) \to \mathrm{Emb}^L(W, M \text{ rel } S)$ *is locally trivial.*

3.2 The Space of Images

As an initial application of these methods, we examine the space of images. This is well-known material (see for example Sect. 44 of [42]), although it seems to be rarely examined in the bounded case. In the next definition, $\mathrm{Diff}(M, V)$ denotes the subgroup of $\mathrm{Diff}(M)$ consisting of the diffeomorphisms that take the submanifold V onto V.

Definition 3.5. Let V be a submanifold of M as in Definition 2.5. The space $\mathrm{Img}(V, M)$ of *images* of V in M is the space of orbits $\mathrm{Diff}(M)/\mathrm{Diff}(M, V)$.

For $j, k \in \mathrm{Diff}(M)$, $j = k$ in $\mathrm{Img}(V, M)$ if and only if $j(V) = k(V)$. Consequently, we may write elements of $\mathrm{Img}(V, M)$ as $j(V)$ with $j \in \mathrm{Diff}(M, V)$.

The next result is basically Theorem 44.1 of [42].

Theorem 3.2. *Let V be a submanifold of a compact manifold M.*

(i) *If V has positive codimension, then* $\mathrm{Img}(V, M)$ *is a Fréchet manifold, locally modeled on the Fréchet space of sections from V to its normal bundle in M. Moreover,* $\mathrm{Diff}(M)$ *is the total space of a locally trivial principal bundle with structure group* $\mathrm{Diff}(M, V)$*, whose base space is* $\mathrm{Img}(V, M)$.

(ii) *If V has codimension zero, and W is the frontier of V in M, then the restriction map* $\mathrm{Img}(V, M) \to \mathrm{Img}(W, M)$ *is either a two-sheeted covering map or a homeomorphism, according to whether or not there exists a diffeomorphism of M that preserves W and interchanges V and $\overline{M - V}$.*

Proof. Assume first that V has positive codimension. The map $\mathrm{Emb}(V, M) \to \mathrm{Img}(V, M)$ is $\mathrm{Diff}(M)$-equivariant, so to prove that $\mathrm{Emb}(V, M) \to \mathrm{Img}(V, M)$ is locally trivial, it suffices to find local $\mathrm{Diff}(M)$ cross-sections at points in $\mathrm{Img}(V, M)$. Since $\mathrm{Diff}(M)$ acts transitively on $\mathrm{Img}(V, M)$, it suffices to find a local cross-section at $1_M\,\mathrm{Diff}(M, V)$.

Let $\nu(V)$ be the normal bundle of V. For some ε, $\mathrm{Exp}\colon \nu_\varepsilon(V) \to M$ is a tubular neighborhood of V, where $\nu_\varepsilon(V)$ is the space of vectors of length less than ε. For each $g\,\mathrm{Diff}(M, V)$ in some neighborhood U of $1_M\,\mathrm{Diff}(M, V)$, $g(V)$ meets each fiber of the tubular neighborhood in a single point. So at each $x \in V$, there is a unique normal vector $X(g)(x)$ in the fiber $\nu_x(V)$ of the normal bundle of V at x such that $\mathrm{Exp}(X(g)(x)) = g(V) \cap \mathrm{Exp}(\nu_x(V))$. This defines $X\colon U \to \mathscr{X}(V, TM)$. Note that X^{-1} defines a local chart for the Fréchet structure of $\mathrm{Img}(V, M)$ at i, showing that $\mathrm{Emb}(V, M)$ is a Fréchet manifold.

By the Extension Lemma 3.2, there is a continuous linear map $k\colon \mathscr{X}(V, TM) \to \mathscr{X}(M, TM)$ such that $k(Y)(x) = Y(x)$ for all x in V and all Y in $\mathscr{X}(V, TM)$. By the Exponentiation Lemma 3.3, there is a neighborhood K of Z in $\mathscr{X}(M, TM)$ such that $\mathrm{TExp}(Y)$ is defined for all $Y \in W$, and TExp carries W into $\mathrm{Diff}(M)$. Provided that our original U was selected small enough that $k(X(U)) \subset K$, the composition $\mathrm{TExp} \circ k \circ X\colon U \to \mathrm{Diff}(M)$ is the desired local cross-section.

Suppose now that V has codimension zero, with frontier W. If there does not exist an element of $\mathrm{Diff}(M, W)$ that interchanges V and $\overline{M - V}$. then $\mathrm{Diff}(M, W) = \mathrm{Diff}(M, V)$, and the restriction map sending $j\,\mathrm{Diff}(M, V)$ to $j|_W\,\mathrm{Diff}(M, W)$ defines a homeomorphism between $\mathrm{Img}(V, M)$ and $\mathrm{Img}(W, M)$. In the remaining case, we fix an element $H_0 \in \mathrm{Diff}(M, W)$ that interchanges V and $\overline{M - V}$.

Define $\varrho\colon \mathrm{Img}(V, M) \to \mathrm{Img}(W, M)$ by sending $j\,\mathrm{Diff}(M, V)$ to $j|_W\,\mathrm{Diff}(M, W)$, i.e. sending $j(V)$ to $j(W)$. This is well-defined, since if $j_1(V) = j_2(V)$ then $j_1(W) = j_2(W)$.

A free involution τ on $\mathrm{Img}(V, M)$ is defined by sending $j(V)$ to $j(H_0(V))$. To see that $\mathrm{Img}(W, M)$ is the quotient of $\mathrm{Img}(V, M)$ by this involution, let $j_1(V), j_2(V) \in \mathrm{Img}(V, M)$ and suppose that $j_1(W) = j_2(W)$. Then either $j_1(V) = j_2(V)$ or $j_1(V) = j_2(\overline{M - V})$. The latter case says exactly that $j_1(V) = j_2(H_0(V))$. □

3.3 Projection of Fiber-Preserving Diffeomorphisms

Throughout this section and the next, it is understood that $p\colon E \to B$ is a locally trivial smooth map as in Sect. 2.6, such that the metric on B is a product near ∂B, and the metric on E is a product near $\partial_h E$ such that the I-fibers of $\partial_h E \times I$ are

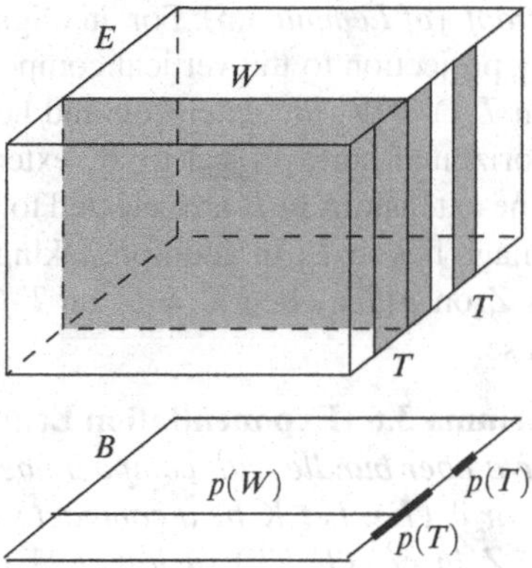

Fig. 3.1 The neighborhood T in Lemma 3.5

vertical. When W is a vertical submanifold of E, it is then automatic that W meets the collar $\partial_h E \times \mathrm{I}$ in I-fibers. By reselecting the metric on B, we may assume that $p(W)$ meets the collar $\partial B \times \mathrm{I}$ in I-fibers. From Definition 2.7, we have the notations $\partial_h W = W \cap \partial_h E$ and $\partial_v W = W \cap \partial_v E$.

We now examine the fundamental lemmas of [51] in the fiber-preserving case. Lemma 3.1 adapts straightforwardly using the aligned exponential from Definition 2.10 and the aligned vector fields from Definition 2.9.

Lemma 3.4 (Logarithm Lemma for fiber-preserving maps). *Assume that $p: E \to B$ has compact fiber, and suppose that W is a compact vertical submanifold of E. Then there are an open neighborhood U of the inclusion i_W in $\mathrm{Emb}_f(W, E)$ and a continuous map $X: U \to \mathscr{A}(W, TE)$ such that for all $j \in U$, $\mathrm{Exp}_a((X(j))(x))$ is defined for all $x \in W$ and $\mathrm{Exp}_a((X(j))(x)) = j(x)$ for all $x \in W$. Moreover, $X(i_W) = Z$.*

Proof. We adapt the argument in Lemma 3.1, using the aligned exponential. Choose ε small enough so that for all $x \in W$, Exp carries the ε-ball about 0 in $T_x(E)$ (that is, the portion of this ε-ball on which it is defined, which may be as small as a closed quarter-ball when $x \in \partial_v E \cap \partial_h E$) diffeomorphically to a neighborhood W_x of x in E. Choose a neighborhood U of i_W in $\mathrm{Emb}_f(W, E)$ so that if $j \in U$ then $j(x) \in W_x$. For $j \in U$ define $X(j)(x)$ to be the unique vector in $T_x(M)$ of length less than ε for which $\mathrm{Exp}_a(X(j)(x))$ equals $j(x)$. Since $p \circ j(x) = p \circ j(y)$ whenever $p(x) = p(y)$, and j is close to the inclusion, $X \in \mathscr{A}(W, TE)$. □

Lemma 3.5 (Extension Lemma for fiber-preserving maps). *Let W be a compact vertical submanifold of E. Let T be a closed fibered neighborhood in $\partial_v E$ of $T \cap \partial_v W$, and let $L \subseteq E$ be a neighborhood of W. Then there is a continuous linear map $k: \mathscr{A}(W, TE) \to \mathscr{A}^L(E, TE)$ such that $k(X)(x) = X(x)$ for all $x \in W$ and all $X \in \mathscr{A}(W, TE)$. If $X(x) = Z(x)$ for all $x \in T \cap \partial_v W$, then $k(X)(x) = Z(x)$ for all $x \in T$. Furthermore, $k(\mathscr{V}(W, TE)) \subset \mathscr{V}^L(E, TE)$.*

Figure 3.1 illustrates the neighborhood T in Lemma 3.5.

Proof (of Lemma 3.5). For an aligned vector field X, we use Lemma 3.2 followed by projection to the vertical components to extend the vertical part. When $X = Z$ on $T \cap \partial_v W$, the extension and hence its vertical projection are Z on T. For the horizontal part, project to B, extend using Lemma 3.2, and take horizontal lifts. The extensions in B are selected to vanish outside a neighborhood L' whose inverse image lies in L. In addition, taking $S = p(T)$ in Lemma 3.2, the extension in B is Z on $p(T)$ when $X = Z$ on $T \cap \partial_v(W)$, ensuring that the lift is Z on T in this case. □

Lemma 3.6 (Exponentiation Lemma for fiber-preserving maps). *Let $p: E \to B$ be a fiber bundle with compact fiber, and assume that the metric on E is a product near $\partial_h(E)$. Let K be a compact subset of E. Then there exists a neighborhood U of Z in $\mathscr{A}^K(E, TE)$ such that* $\mathrm{TExp}_a(X(x))$ *is defined for all $X \in U$ and* TExp_a *carries U into* $\mathrm{Diff}_f^K(E)$.

Proof. As in the proof of Lemma 3.3, enlarge E to a complete open manifold N by attaching $\partial_h E \times [0, \infty)$ along $\partial_h E$. For each fiber F in E, put $\partial_h F = F \cap \partial_h E$. Then N is still a fiber bundle over B, where each fiber F has been enlarged to an open manifold by attaching $\partial_h F \times [0, \infty)$ along $F \cap \partial_h E$. We denote this fibering by $p_N: N \to B$.

Now, consider a vector field $X \in \mathscr{A}^K(E, TE)$. We extend the vertical and horizontal parts X_v and X_h to N separately. For X_v, we extend using Lemma 2.1, then project into the vertical subspace at each point. For X_n, at each point $x \in N - E$, we just take the horizontal lift of $p_*(X(y))$ for some $y \in E$ with $p(y) = p_N(x)$, so that the extended vector field is aligned. Its restriction to $E \cup \partial_h E \times [0, 1]$ is tangent to $\partial_h E$. The vertical part of the extension, constructed using Lemma 2.1, vanishes off of $K \cup (K \cap \partial_h E) \times [0, 1]$. The proof is now completed as in Lemma 3.3, using the aligned exponential on $E \cup \partial_h E \times [0, 1]$, and taking $L = K \cup (K \cap \partial_h E) \times [0, 1]$. □

Definition 3.6. Let $p: E \to B$ be a fiber bundle. For an element g of $\mathrm{Diff}_f(E)$, the induced diffeomorphism of B will be denoted by $\overline{g}$. More generally, if W is a vertical submanifold of E, each $j \in \mathrm{Emb}_f(W, E)$ induces an embedding of $p(W)$ into B, denoted by $\overline{j}$. Note that $\mathrm{Diff}_f(E)$ acts on $\mathrm{Diff}(B)$ and on $\mathrm{Emb}(p(W), B)$ by sending h to $\overline{g}h$. More generally, if $S \subset \partial B$ is a neighborhood of $S \cap \partial\, p(W)$ and K is a neighborhood of W, then $\mathrm{Diff}_f^{p^{-1}(K)}(E \text{ rel } p^{-1}(S))$ acts in this way on $\mathrm{Diff}^K(B \text{ rel } S)$ and $\mathrm{Emb}^K(W, B \text{ rel } S)$.

Theorem 3.3. *Let K be a compact subset of B, let S be a subset of ∂B, and put $T = p^{-1}(S)$. Then* $\mathrm{Diff}^K(B \text{ rel } S)$ *admits local* $\mathrm{Diff}_f^{p^{-1}(K)}(E \text{ rel } T)$ *cross-sections.*

Proof. By Proposition 3.2, it suffices to find a local cross-section at the identity id_B. Let L be a compact codimension-zero submanifold of B that contains K in its topological interior (and such that as usual, L meets $\partial B \times I$ in I-fibers). By Lemma 3.1, there are a neighborhood U of the inclusion i_L in $\mathrm{Emb}(L, B)$ and a continuous map $X: U \to \mathscr{X}(L, TB)$ such that for all $j \in U$ and all $x \in L$, $\mathrm{Exp}(X(j)(x))$ is defined and $\mathrm{Exp}(X(j)(x)) = j(x)$.

Suppose that $f \in \mathrm{Diff}^K(B \text{ rel } S)$. Then $X(f|_L)$ vanishes on a neighborhood in L of the frontier of L in B, and on $L \cap S$, so the vector field $X(f|_L)$ extends to a smooth vector field $X'(f|_L)$ on B using Z on $B - K$, which vanishes on S. For each $x \in B$, $\mathrm{Exp}(X'(j)(x))$ is defined and $\mathrm{Exp}(X'(j)(x)) = f(x)$. At each point y of E, let $\widetilde{X'}(j)(y)$ be the horizontal lift of $X'(j)(p(y))$. This produces an aligned vector field in $\mathscr{A}^{p^{-1}(L)}(E, TE)$, which vanishes on T.

Choose a neighborhood V of id_B in $\mathrm{Diff}^K(B \text{ rel } S)$ such that $f|_L \in U$ for each $f \in V$. On V, define $\chi(f) = \mathrm{TExp}_a(\widetilde{X'}(f|_L))$. Since $\widetilde{X'}(f|_L))$ vanishes on T and off of $p^{-1}(K)$, this defines $\chi\colon V \to \mathrm{Diff}_f^{p^{-1}(K)}(E \text{ rel } T)$. This is a local cross-section, since given $b \in K \subseteq L$ we may choose any y with $p(y) = b$ and calculate that for the induced diffeomorphism $\overline{\chi(f)}$ on B,

$$\begin{aligned}\overline{\chi(f)}(b) &= p(\chi(f)(x)) \\ &= p\big(\mathrm{Exp}_a\big(\widetilde{X}(\varrho(f))(x)\big)\big) \\ &= \mathrm{Exp}\big(X(\varrho(f))(b)\big) \\ &= f(b)\end{aligned}$$

while at points in $B - K$, $\overline{\chi(f)}(b) = b$. □

From Proposition 3.1, we have immediately

Theorem 3.4. *Let K be a compact subset of B. Let $S \subseteq \partial B$ and let $T = p^{-1}(S)$. Then $\mathrm{Diff}_f^{p^{-1}(K)}(E \text{ rel } T) \to \mathrm{Diff}^K(B \text{ rel } S)$ is locally trivial.*

Each of the fibration theorems we prove has a corresponding corollary involving parameterized lifting or extension, but since the statements are all analogous we give only the following one as a prototype.

Corollary 3.3. *(Parameterized Isotopy Lifting Theorem) Let K be a compact subset of B, let $S \subseteq \partial B$, and let $T = p^{-1}(S)$. Suppose that for each t in a path-connected parameter space P there is an isotopy $g_{t,s}$, such that each $g_{t,s}$ is the identity on S and outside of K, and such that $g_{t,0}$ lifts to a diffeomorphism $G_{t,0}$ of E which is the identity on T. Assume that sending $(t,s) \to g_{t,s}$ defines a continuous function from $P \times [0,1]$ to* Diff(B rel S) *and sending t to $G_{t,0}$ defines a continuous function from P to* Diff(E rel T). *Then the family $G_{t,0}$ extends to a continuous family on $P \times$*I *such that for each (t,s), $G_{t,s}$ is a fiber-preserving diffeomorphism inducing $g_{t,s}$ on B.*

3.4 Restriction of Fiber-Preserving Diffeomorphisms

In this section we present the analogues of the main results of Palais [51] in the fibered case. As usual, we tacitly assume that metrics are products near the boundary and that submanifolds meet the boundary in I-fibers. We remind the reader about Fig. 3.1, which indicates the setup in the next result.

Theorem 3.5. *Let W be a compact vertical submanifold of E. Let T be a closed fibered neighborhood in $\partial_v E$ of $T \cap \partial_v W$, and let L be a neighborhood of W. Then*

(i) $\mathrm{Emb}_f^L(W, E \text{ rel } T)$ *admits local* $\mathrm{Diff}_f^L(E \text{ rel } T)$ *cross-sections.*
(ii) $\mathrm{Emb}_v(W, E \text{ rel } T)$ *admits local* $\mathrm{Diff}_v^L(E \text{ rel } T)$ *cross-sections.*

Proof. By Proposition 3.2, it suffices to find local cross-sections at the inclusion i_W. Choose a compact neighborhood K of W with $K \subseteq L$ and $K = p^{-1}(p(K))$.

By Lemma 3.4, there are a neighborhood U_1 of the inclusion i_W in $\mathrm{Emb}_f(W, E)$ and a continuous map $X: U_1 \to \mathscr{A}(W, TE)$ such that for all $j \in U_1$ and all $x \in W$, $\mathrm{Exp}_a(X(j)(x))$ is defined and $\mathrm{Exp}_a(X(j)(x)) = j(x)$. By Lemma 3.5, there is a continuous linear map $k: \mathscr{A}(W, TE) \to \mathscr{A}^K(E, TE)$, with $k(\mathscr{V}(W, TE)) \subset \mathscr{V}^K(E, TE))$, such that $k(X)(x) = X(x)$ for all $x \in W$. Lemma 3.6 now gives a neighborhood U_2 of Z in $\mathscr{A}^K(E, TE)$ such that $\mathrm{Exp}_a(X)$ is defined for all $X \in U_2$, and TExp_a has image in $\mathrm{Diff}_f^K(E)$. Putting $U = X^{-1}(k^{-1}(U_2))$, the composition $\mathrm{TExp}_a \circ k \circ X: U \to \mathrm{Diff}_f^K(E)$ is the desired cross-section for (i).

Since X carries $\mathrm{Emb}_v(W, B)$ into $\mathscr{V}(W, TE)$, k carries $\mathscr{V}(W, TE)$ into $\mathscr{V}^K(E, TE)$, and TExp_a carries $U_2 \cap \mathscr{V}^K(E, TE)$ into $\mathrm{Diff}_v(E)$, this cross-section restricts on $\mathrm{Emb}_v(W, E \text{ rel } T)$ to a $\mathrm{Diff}_v^L(E \text{ rel } T)$ cross-section, giving (ii). □

Proposition 3.1 has the following immediate corollaries.

Corollary 3.4. *Let W be a compact vertical submanifold of E. Let T be a closed fibered neighborhood in $\partial_v E$ of $T \cap \partial_v W$, and L a neighborhood of W. Then the following restrictions are locally trivial:*

(i) $\mathrm{Diff}_f^L(E \text{ rel } T) \to \mathrm{Emb}_f^L(W, E \text{ rel } T)$.
(ii) $\mathrm{Diff}_v^L(E \text{ rel } T) \to \mathrm{Emb}_v(W, E \text{ rel } T)$.

Corollary 3.5. *Let V and W be vertical submanifolds of E, with $W \subseteq V$. Let T be a closed fibered neighborhood in $\partial_v E$ of $T \cap \partial_v V$, and let L a neighborhood of V. Then the following restrictions are locally trivial:*

(i) $\mathrm{Emb}_f^L(V, E \text{ rel } T) \to \mathrm{Emb}_f^L(W, E \text{ rel } T)$.
(ii) $\mathrm{Emb}_v(V, E \text{ rel } T) \to \mathrm{Emb}_v(W, E \text{ rel } T)$.

The final result of this section is the projection-restriction square for bundles.

Theorem 3.6. *Let W be a compact vertical submanifold of E. Let K be a compact neighborhood of $p(W)$ in B. Let T be a closed fibered neighborhood in $\partial_v E$ of $T \cap \partial_v W$, and put $S = p(T)$. Then all four maps in the following commutative square are locally trivial:*

$$\begin{array}{ccc} \mathrm{Diff}_f^{p^{-1}(K)}(E \text{ rel } T) & \longrightarrow & \mathrm{Emb}_f^{p^{-1}(K)}(W, E \text{ rel } T) \\ \downarrow & & \downarrow \\ \mathrm{Diff}^K(B \text{ rel } S) & \longrightarrow & \mathrm{Emb}^K(p(W), B \text{ rel } S)\,. \end{array}$$

Proof. The top arrow is Corollary 3.4(i), the left vertical arrow is Theorem 3.3, and the bottom arrow is Corollary 3.1. For the right vertical arrow, we will first show that $\mathrm{Emb}^K(p(W), B \text{ rel } S)$ admits local $\mathrm{Diff}_f^{p^{-1}(K)}(E \text{ rel } T)$ cross-sections. Let $i \in \mathrm{Emb}^K(p(W), B \text{ rel } S)$. Using Theorems 3.5 and 3.3, choose local cross-sections $\chi_1 \colon U \to \mathrm{Diff}^K(B \text{ rel } S)$ at i and $\chi_2 \colon V \to \mathrm{Diff}_f^{p^{-1}(K)}(E \text{ rel } T)$ at $\chi_1(i)$. Let $U_1 = \chi_1^{-1}(V)$, then for $j \in U_1$ we have

$$\overline{\chi_2\chi_1(j)}i = \overline{\chi_2(\chi_1(j))}i = \chi_1(j)i = j \, .$$

Since the right vertical arrow is $\mathrm{Diff}_f^{p^{-1}(K)}(E \text{ rel } T)$-equivariant, Proposition 3.1 implies that it is locally trivial. □

3.5 Restriction Theorems for Orbifolds

Throughout this section, indeed in all of our work, an orbifold means an orbifold in the standard sense whose universal covering $\pi \colon \widetilde{\mathcal{O}} \to \mathcal{O}$ is a manifold. We assume further that $\mathcal{O}$ is a smooth orbifold, meaning that $\widetilde{\mathcal{O}}$ is a smooth manifold and the group H of covering transformations consists of diffeomorphisms.

Definition 3.7. A map $f \colon \widetilde{\mathcal{O}} \to \widetilde{\mathcal{O}}$ is called *(weakly) H-equivariant* if for some automorphism α of H, $f(h(x)) = \alpha(h)(f(x))$ for all $x \in \widetilde{\mathcal{O}}$ and $h \in H$. Define $C^\infty_H(\widetilde{\mathcal{O}})$ to be the space of H-equivariant boundary-preserving smooth maps from $\widetilde{O}$ to $\widetilde{O}$, and $\mathrm{Diff}_H(\widetilde{\mathcal{O}})$ to be the H-equivariant diffeomorphisms of $\widetilde{\mathcal{O}}$. Note that $\mathrm{Diff}_H(\widetilde{\mathcal{O}})$ is the normalizer of H in $\mathrm{Diff}(\widetilde{\mathcal{O}})$.

Definition 3.8. An *orbifold homeomorphism* of $\mathcal{O}$ is a homeomorphism of the underlying topological space of $\mathcal{O}$ that is induced by an H-equivariant homeomorphism of $\widetilde{\mathcal{O}}$, called a *lift* of the orbifold homeomorphism. An orbifold diffeomorphism of $\mathcal{O}$ is an orbifold homeomorphism for which some and hence all lifts to $\widetilde{\mathcal{O}}$ are diffeomorphisms. Define $\mathrm{Diff}(\mathcal{O})$ to be the group of orbifold diffeomorphisms. Note that $\mathrm{Diff}(\mathcal{O})$ is the quotient of $\mathrm{Diff}_H(\widetilde{\mathcal{O}})$ by the normal subgroup H. We give $\mathrm{Diff}(\mathcal{O})$ the quotient topology of the C^∞-topology on $\mathrm{Diff}_H(\widetilde{\mathcal{O}})$.

Definition 3.9. An orbifold $\mathcal{W}$ contained in $\mathcal{O}$ is called a *suborbifold* of $\mathcal{O}$ if its inverse image $\widetilde{\mathcal{W}}$ in $\widetilde{\mathcal{O}}$ is a submanifold. An element of $\mathrm{Emb}(\widetilde{\mathcal{W}}, \widetilde{\mathcal{O}})$ is called *H-equivariant* if it extends to an element of $\mathrm{Diff}_H(\widetilde{\mathcal{O}})$, and the subspace of H-equivariant embeddings is denoted by $\mathrm{Emb}_H(\widetilde{\mathcal{W}}, \widetilde{\mathcal{O}})$. An *embedding* of $\mathcal{W}$ into $\mathcal{O}$ is an embedding induced by an element of $\mathrm{Emb}_H(\widetilde{\mathcal{W}}, \widetilde{\mathcal{O}})$, and the space of embeddings is denoted by $\mathrm{Emb}(\mathcal{W}, \mathcal{O})$.

Throughout this section, $\mathcal{W}$ will denote a compact suborbifold of $\mathcal{O}$.

Definition 3.10. A section from an H-equivariant subset $\widetilde{L}$ of $\widetilde{\mathcal{O}}$ to $T\widetilde{\mathcal{O}}|_{\widetilde{L}}$ is called *H-equivariant* if for each $x \in \widetilde{L}$ and each $h \in H$, $h_*(X(x)) = X(h(x))$. In general, we use a subscript H to indicate the H-equivariant elements of any of the spaces of sections that we have defined, thus for example $\mathscr{X}_H(\widetilde{\mathscr{W}}, T\widetilde{\mathscr{O}})$ means the H-equivariant elements of $\mathscr{X}(\widetilde{\mathscr{W}}, T\widetilde{\mathscr{O}})$.

The next two lemmas provide equivariant functions and metrics.

Lemma 3.7. *Let H be a group acting smoothly and properly discontinuously on a manifold M, possibly with boundary, such that M/H is compact. Let A be an H-invariant closed subset of M, and U an H-invariant neighborhood of A. Then there exists an H-equivariant smooth function $\gamma\colon M \to [0,1]$ that is identically equal to 1 on A and whose support is contained in U.*

Proof. Fix a compact subset C of M which maps surjectively onto M/H under the quotient map. Let $\phi\colon M \to [0,\infty)$ be a smooth function such that $\phi(x) \geq 1$ for all $x \in C \cap A$ and whose support is compact and contained in U. Define ψ by $\psi(x) = \sum_{h\in H} \phi(h(x))$. Now choose $\eta\colon \mathbb{R} \to [0,1]$ such that $\eta(r) = 0$ for $r \leq 0$ and $\eta(r) = 1$ for $r \geq 1$, and put $\gamma = \eta \circ \psi$. □

When $\mathscr{O}$ is compact, the following lemma provides a Riemannian metric on $\widetilde{\mathscr{O}}$ for which the covering transformations are isometries.

Lemma 3.8. *Let H be a group acting smoothly and properly discontinuously on a manifold M, possibly with boundary, such that M/H is compact. Let N be a properly embedded H-invariant submanifold, possibly empty. Then M admits a complete H-equivariant Riemannian metric, which is a product near ∂M, and such that N meets $\partial M \times \mathrm{I}$ in I-fibers. Moreover, the action preserves the collar, and if $(y,t) \in \partial M \times \mathrm{I}$ and $h \in H$, then $h(y,t) = (h|_{\partial M}(y), t)$.*

Proof. We first prove that equivariant Riemannian metrics exist. Choose a compact subset C of M that maps surjectively onto M/H under the quotient map. Let $\phi\colon M \to [0,\infty)$ be a compactly supported smooth function which is positive on C. Choose a Riemannian metric R on M and denote by R_x the inner product which R assigns to $T_x(M)$. Define a new metric R' by

$$R'_x(v,w) = \sum_{h\in H} \phi(h(x))\, R_{h(x)}(h_*(v), h_*(w))\ .$$

Since ϕ is compactly supported, the sum is finite, and since every orbit meets the support of ϕ, R' is positive definite. To check equivariance, let $g \in H$. Then

$$\begin{aligned} R'_{g(x)}(g_*(v), g_*(w)) &= \sum_{h\in H} \phi(h(g(x)))\, R_{h(g(x))}(h_*(g_*(v)), h_*(g_*(w))) \\ &= \sum_{h\in H} \phi(hg(x))\, R_{hg(x)}((hg)_*(v), (hg)_*(w)) \\ &= R'_x(v,w)\ . \end{aligned}$$

We need to improve the metric near the boundary. First, note that $C \cap \partial M$ maps surjectively onto the image of ∂M. Choose an inward-pointing vector field τ' on a neighborhood U of $C \cap \partial M$, which is tangent to N. Choose a smooth function $\phi\colon M \to [0,\infty)$ which is positive on $C \cap \partial M$ and has compact support contained in U. The field $\phi\tau'$ defined on U extends using the zero vector field on $M - U$ to a vector field τ which is nonvanishing on $C \cap \partial M$. For x in the union of the H-translates of U, define $\omega_x = \sum_{h\in H} \phi(h(x))\, h_*^{-1}(\tau_{h(x)})$. This is defined, nonsingular, and equivariant on an H-invariant neighborhood of ∂M, and we use it to define a collar $\partial M \times [0,2]$ equivariant in the sense that if $(y,t) \in \partial M \times [0,2]$ then $h(y,t) = (h|_{\partial M}(y), t)$. Moreover, N meets this collar in I-fibers. On $\partial M \times [0,2]$, choose an equivariant metric R_1 which is the product of an equivariant metric on ∂M and the standard metric on $[0,2]$, and choose any equivariant metric R_2 defined on all of M. Using Lemma 3.7, choose H-equivariant functions ϕ_1 and ϕ_2 from M to $[0,1]$ so that $\phi_1(x) = 1$ for all $x \in \partial M \times [0,3/2]$ and the support of ϕ_1 is contained in $\partial M \times [0,2)$, and so that $\phi_2(x) = 1$ for $x \in M - \partial M \times [0,3/2]$ and the support of ϕ_2 is contained in $M - \partial M \times [0,1]$. Then, $\phi_1 R_1 + \phi_2 R_2$ is H-equivariant and is a product near ∂M, and N is vertical in $\partial M \times \mathrm{I}$.

Since M/H is compact and H acts as isometries, the metric must be complete. For let C be a compact subset of M that maps surjectively onto M/H. We may enlarge C to a compact codimension-zero submanifold C' such that every point of M has a translate which lies in C' at distance at least a fixed ε from the frontier of C'. Then, any Cauchy sequence in M can be translated, except for finitely many terms, into a Cauchy sequence in C'. Since C' is compact, this converges, so the original sequence also converged. □

Proposition 3.3. *Suppose that H acts properly discontinuously on a locally compact connected Hausdorff space X, and that X/H is compact. Then H is finitely generated.*

Proof. Using local compactness, there exists a compact set C whose interior maps surjectively to X/H. Let H_0 be the subgroup generated by the finitely many elements h such that $h(C) \cap C$ is nonempty. The union of the H_0-translates of C is an open and closed subset, so must equal X. This implies that $H = H_0$. □

Definition 3.11. Let A be an H-invariant subset of $\widetilde{O}$. Define $(\mathrm{C}^\infty)_H^A(\widetilde{\mathscr{O}})$ to be the elements of $\mathrm{C}^\infty_{\mathrm{H}}(\widetilde{\mathscr{O}})$ that fix each point not in A, and define $\mathrm{Diff}_H^A(\widetilde{\mathscr{O}})$ similarly. If A is a neighborhood of $\widetilde{\mathscr{W}}$, define $\mathrm{Emb}_H^A(\widetilde{\mathscr{W}}, \widetilde{\mathscr{O}})$ to be the elements of $\mathrm{Emb}_H(\widetilde{\mathscr{W}}, \widetilde{\mathscr{O}})$ that carry $\widetilde{W}$ into A. We use this notation to extend our previous concepts to orbifolds. For example, if K is a neighborhood of a suborbifold $\mathscr{W}$ in $\mathscr{O}$, then $\mathrm{Emb}^K(\mathscr{W}, \mathscr{O})$ is the subspace of $\mathrm{Emb}(\mathscr{W}, \mathscr{O})$ induced by elements of $\mathrm{Emb}_H^{\pi^{-1}(K)}(\widetilde{\mathscr{W}}, \widetilde{\mathscr{O}})$, $\mathscr{X}^K(\mathscr{O}, T\mathscr{O})$ is the subspace of elements of $\mathscr{X}_H^K(\mathscr{O}, T\mathscr{O})$ that equal Z outside of $\pi^{-1}(K)$, and so on.

Lemma 3.9. *Suppose that H acts properly discontinuously as isometries on $\widetilde{\mathscr{O}}$. Let $\widetilde{K}$ be an H-invariant subset of $\widetilde{\mathscr{O}}$ whose quotient in $\mathscr{O}$ is compact. Then there exists a neighborhood J of $1_{\widetilde{\mathscr{O}}}$ in $(\mathrm{C}^\infty)_H^{\widetilde{K}}(\widetilde{\mathscr{O}})$ that consists of diffeomorphisms.*

Proof. Assume for now that $\mathcal{O}$ is compact and $\widetilde{\mathcal{O}} = \widetilde{K}$, and fix a compact set C in $\widetilde{\mathcal{O}}$ that maps surjectively to $\mathcal{O}$.

We claim that if $f \in \mathrm{C}^{\infty}_{\mathrm{H}}(\widetilde{\mathcal{O}})$ is close enough to $1_{\widetilde{\mathcal{O}}}$, then f commutes with the H-action. By Proposition 3.3, H is finitely generated. Choose an $x \in \widetilde{\mathcal{O}}$ which is not fixed by any nontrivial element of H. Define $\Phi\colon (\mathrm{C}^{\infty})^{\widetilde{K}}_{H}(\widetilde{\mathcal{O}}) \to \mathrm{End}(H)$ by sending f to ϕ_f where $f(h(x)) = \phi_f(h)f(x)$. This is independent of the choice of x, and is a homomorphism. If f is close enough to $1_{\widetilde{\mathcal{O}}}$ on $\{x, h_1(x), \ldots, h_n(x)\}$, where $\{h_1, \ldots, h_n\}$ generates H, then $\phi_f = 1_H$. This prove the claim.

For the remainder of the argument, we require f to be close enough to $1_{\widetilde{\mathcal{O}}}$ to ensure that f commutes with the H-action. This implies that $f^{-1}(S)$ is compact whenever S is compact. For if S is a subset for which $f^{-1}(S)$ meets infinitely many translates of C, then S meets infinitely many translates of $f(C)$, so S cannot be compact.

Requiring in addition that f be sufficiently C^{∞}-close to $1_{\widetilde{\mathcal{O}}}$, we have f_* nonsingular at each point of C, hence on all of $\widetilde{\mathcal{O}}$. Since f takes boundary to boundary, it follows that f is a local diffeomorphism. Since inverse images of compact sets under f are compact, f is a covering map. And since $\widetilde{\mathcal{O}}$ is simply-connected, f is a diffeomorphism.

Now suppose that $\mathcal{O}$ is noncompact. Choose a compact codimension-zero suborbifold $\mathcal{L}$ of $\mathcal{O}$ that contains $\widetilde{K}/H$ in its topological interior. Each element of $(\mathrm{C}^{\infty})^{\widetilde{K}}_{H}(\widetilde{\mathcal{L}})$ extends to an element of $(\mathrm{C}^{\infty})^{\widetilde{K}}_{H}(\widetilde{\mathcal{O}})$ by using the identity on $\widetilde{\mathcal{O}} - \widetilde{\mathcal{L}}$. Applying the case when $\mathcal{O}$ is compact, that is, using $\mathcal{L}$ in place of $\mathcal{O}$, some neighborhood of the identity in $\mathrm{C}^{\infty}_{\mathrm{H}}(\widetilde{\mathcal{L}})$ consists of maps which are diffeomorphisms on $\widetilde{\mathcal{L}}$. The intersection of this neighborhood with $(\mathrm{C}^{\infty})^{\widetilde{K}}_{H}(\widetilde{\mathcal{L}})$ consists of diffeomorphisms, and their extensions to $\widetilde{\mathcal{O}}$ form the desired neighborhood of the identity in $(\mathrm{C}^{\infty})^{\widetilde{K}}_{H}(\widetilde{\mathcal{O}})$. □

We now prove the analogues of Lemmas 3.1 and 3.2 for vector fields on $\mathcal{O}$. Assume that $\mathcal{W}$ is a compact suborbifold of $\mathcal{O}$.

Lemma 3.10 (Equivariant Logarithm Lemma). *There are a neighborhood U of the inclusion $i_{\widetilde{\mathcal{W}}}$ of $\widetilde{\mathcal{W}}$ into $\widetilde{\mathcal{O}}$ in $\mathrm{Emb}_H(\widetilde{\mathcal{W}}, \widetilde{\mathcal{O}})$ and a continuous map $X\colon U \to \mathcal{X}_H(\widetilde{\mathcal{W}}, T\widetilde{\mathcal{O}})$ such that for all $j \in U$, $\mathrm{Exp}(X(j)(x))$ is defined for all $x \in \widetilde{\mathcal{W}}$ and $\mathrm{Exp}(X(j)(x)) = j(x)$ for all $x \in \widetilde{\mathcal{W}}$. Moreover, $X(i_{\widetilde{\mathcal{W}}}) = Z$.*

Proof. By replacing $\mathcal{O}$ by a compact orbifold neighborhood of $\mathcal{W}$ and applying Lemma 3.8, we may assume that H acts as isometries on $\widetilde{\mathcal{O}}$, that the metric is a product near $\partial\widetilde{\mathcal{O}}$, and that $\widetilde{\mathcal{W}}$ meets the collar $\partial\widetilde{\mathcal{O}} \times \mathrm{I}$ in I-fibers. The proof then follows the argument of Lemma 3.1, working equivariantly in $\widetilde{\mathcal{O}}$. □

Lemma 3.11 (Equivariant Extension Lemma). *Let $\mathcal{W}$ be a compact suborbifold of $\mathcal{O}$. Let L be a neighborhood of $\mathcal{W}$ in $\mathcal{O}$ and let S be a closed neighborhood in $\partial\mathcal{O}$ of $S \cap \partial\mathcal{W}$. Denote the inverse images in $\widetilde{\mathcal{O}}$ by $\widetilde{L}$ and $\widetilde{S}$. Then there exists a continuous map $k\colon \mathcal{X}_H(\widetilde{\mathcal{W}}, T\widetilde{\mathcal{O}}) \to \mathcal{X}^{\widetilde{L}}_H(\widetilde{\mathcal{O}}, T\widetilde{\mathcal{O}})$ such that $k(X)(x) = X(x)$ for all x in $\widetilde{\mathcal{W}}$. Moreover, $k(Z) = Z$, and if $X(x) = Z(x)$ for all $x \in \widetilde{S} \cap \partial\widetilde{\mathcal{W}}$, then $k(X)(x) = Z(x)$ for all $x \in \widetilde{S}$.*

Proof. Assume first that $\mathscr{W}$ has positive codimension. Replacing $\mathscr{O}$ by a compact orbifold neighborhood $\mathscr{O}'$ of $\mathscr{W}$, L by a compact neighborhood of $\mathscr{W}$ in $L \cap \mathscr{O}'$, and S by $S \cap \mathscr{O}'$, and using Lemma 3.8, we may assume that H acts as isometries on $\widetilde{\mathscr{O}}$, that the metric is a product near $\partial\widetilde{\mathscr{O}}$, and that $\widetilde{\mathscr{W}}$ meets the collar $\partial\widetilde{\mathscr{O}} \times \mathrm{I}$ in I-fibers. Let $\nu(\widetilde{\mathscr{W}})$ be the normal bundle, regarded as a subbundle of the restriction of $T\widetilde{\mathscr{O}}$ to $\widetilde{\mathscr{W}}$. For $\varepsilon > 0$, let $\nu_\varepsilon(\widetilde{\mathscr{W}})$ be the subspace of all vectors of length less than ε. Since $\widetilde{\mathscr{W}}$ is compact and H acts as isometries on $\widetilde{L}$, Exp embeds $\nu_\varepsilon(\widetilde{\mathscr{W}})$ as a tubular neighborhood of $\widetilde{\mathscr{W}}$ for sufficiently small ε. By choosing ε small enough, we may assume that $\mathrm{Exp}(\nu_\varepsilon(\widetilde{\mathscr{W}})) \subset \widetilde{L}$, that the fibers at points in $\widetilde{S}$ map into $\widetilde{S}$, and that the fibers at points in $\partial\widetilde{\mathscr{O}} - \widetilde{S}$ map into $\partial\widetilde{\mathscr{O}} - \widetilde{S}$.

Now use Lemma 3.7 to choose an H-equivariant smooth function $\alpha\colon \widetilde{\mathscr{O}} \to [0,1]$ which is identically equal to 1 on $\widetilde{\mathscr{W}}$ and has support in $\mathrm{Exp}(\nu_{\varepsilon/2}(\widetilde{\mathscr{W}}))$. The extension $k(X)$ can now be defined exactly as in Lemma 3.2. Note that since H acts as isometries, the parallel translation function P is H-equivariant, and the H-equivariance of $k(X)$ follows easily.

Assume now that $\mathscr{W}$ has codimension zero. The frontier W of $\widetilde{\mathscr{W}}$ is an equivariant properly embedded submanifold of $\widetilde{\mathscr{O}}$. Since H acts as isometries, we can select an equivariant tubular neighborhood of W parameterized as $W \times (-\infty, \infty)$ with $\widetilde{\mathscr{W}} \cap (W \times (-\infty,\infty)) = W \times [0,\infty)$, and so that the action of H respects the $(-\infty,\infty)$-coordinate. By Lemma 2.1, there is a continuous linear extension operator carrying each vector field on $\widetilde{\mathscr{W}}$ to a vector field on $\widetilde{\mathscr{W}} \cup (W \times (-\infty,\infty))$. The extended vector fields are equivariant since they are defined by a formula in terms of the coordinates of $W \times [0,\infty)$. At points of $\partial\widetilde{\mathscr{O}}$, the component of each vector in the direction perpendicular to $\partial\widetilde{\mathscr{O}}$ is 0, so the extended component is also 0 and therefore the extended vector fields are also tangent to the boundary. After multiplying by an equivariant function on $\widetilde{\mathscr{W}} \cup (W \times (-\infty,\infty))$ that is 1 on $\widetilde{\mathscr{W}}$ and 0 on $W \times (-\infty, -1]$, these vector fields extend using Z on $\widetilde{\mathscr{O}} - (\widetilde{\mathscr{W}} \cup (W \times (-\infty,\infty)))$.

□

Now we are ready for the analogue of Theorem B of [51]. Its statement and proof use some notation explained in Definition 3.11.

Theorem 3.7. *Let $\mathscr{W}$ be a compact suborbifold of $\mathscr{O}$. Let S be a closed neighborhood in $\partial\mathscr{O}$ of $S \cap \partial\mathscr{W}$, and let L be a neighborhood of $\mathscr{W}$ in $\mathscr{O}$. Then* $\mathrm{Emb}^L(\mathscr{W}, \mathscr{O} \text{ rel } S)$ *admits local* $\mathrm{Diff}^L(\mathscr{O} \text{ rel } S)$ *cross-sections.*

Proof. By Proposition 3.2, it suffices to find a local cross-section at the inclusion $i_{\mathscr{W}}$. Choose a compact neighborhood K of $\mathscr{W}$ with $K \subseteq L$. Using Lemmas 3.10 and 3.11, we obtain an open neighborhood $\widetilde{V}$ of $i_{\widetilde{\mathscr{W}}}$ in $\mathrm{Emb}_H(\widetilde{\mathscr{W}}, \widetilde{\mathscr{O}})$ and continuous maps $X\colon \widetilde{V} \to \mathscr{X}_H(\widetilde{\mathscr{W}}, T\widetilde{\mathscr{O}})$ and $k\colon \mathscr{X}_H(\widetilde{\mathscr{W}}, T\widetilde{\mathscr{O}}) \to \mathscr{X}_H^{\widetilde{L}}(\widetilde{O}, T\widetilde{\mathscr{O}})$. By Lemma 3.9, there is a neighborhood J of $1_{\widetilde{\mathscr{O}}}$ in $(\mathrm{C}^\infty)_H^{\widetilde{K}}(\widetilde{\mathscr{O}})$ that consists of diffeomorphisms.

On a sufficiently small neighborhood $\widetilde{U}$ of $i_{\widetilde{\mathscr{W}}}$, the function $\widetilde{\chi}\colon \widetilde{U} \to \mathrm{Diff}_H^{\widetilde{K}}(\widetilde{\mathscr{O}})$ defined by $\widetilde{\chi}(j) = \mathrm{TExp} \circ k \circ X(j)$ is defined and has image in J. Let U be the embeddings of $\mathscr{W}$ in $\mathscr{O}$ which admit a lift to $\widetilde{U}$. By choosing $\widetilde{U}$ small enough, we may ensure that the lift of an element of U is unique. Define $\chi\colon U \to \mathrm{Diff}^K(\mathscr{O})$

to be $\widetilde{\chi}$ applied to the lift of an element of U to $\widetilde{U}$, followed by the projection of $\mathrm{Diff}^K_H(\widetilde{\mathcal{O}})$ to $\mathrm{Diff}^K(\mathcal{O})$.

For elements in $U \cap \mathrm{Emb}^K(\mathcal{W}, \mathcal{O} \text{ rel } S)$, each lift to $\widetilde{U}$ that is sufficiently close to $i_{\widetilde{\mathcal{W}}}$ must agree with $i_{\widetilde{\mathcal{W}}}$ on $\widetilde{S}$. So U may be chosen small enough so that if $j \in U$ then its lift $\widetilde{j}$ in $\widetilde{U}$ lies in $\mathrm{Emb}(\widetilde{\mathcal{W}}, \widetilde{\mathcal{O}} \text{ rel } \widetilde{S})$. Then, $X(\widetilde{j}(x)) = Z(x)$ for all $x \in \widetilde{S}$, so $k(X)(x) = Z(x)$ for all $x \in \widetilde{S}$. It follows that $\chi(j) \in \mathrm{Diff}(\mathcal{O} \text{ rel } S)$. □

Corollary 3.6. *Let $\mathcal{W}$ be a compact suborbifold of $\mathcal{O}$, which is either properly embedded or codimension-zero. Let S be a closed neighborhood in $\partial\,\mathcal{O}$ of $S \cap \partial\mathcal{W}$, and let L be a neighborhood of $\mathcal{W}$ in $\mathcal{O}$. Then the restriction* $\mathrm{Diff}^L(\mathcal{O} \text{ rel } S) \to \mathrm{Emb}^L(\mathcal{W}, \mathcal{O} \text{ rel } S)$ *is locally trivial.*

Corollary 3.7. *Let $\mathcal{V}$ and $\mathcal{W}$ be suborbifolds of $\mathcal{O}$, with $\mathcal{W} \subset \mathcal{V}$. Assume that $\mathcal{W}$ compact, and is either properly embedded or codimension-zero. Let S be a closed neighborhood in $\partial\,\mathcal{O}$ of $S \cap \partial\mathcal{W}$, and let L be a neighborhood of $\mathcal{W}$ in $\mathcal{O}$. Then the restriction* $\mathrm{Emb}^L(\mathcal{V}, \mathcal{O} \text{ rel } S) \to \mathrm{Emb}^L(\mathcal{W}, \mathcal{O} \text{ rel } S)$ *is locally trivial.*

3.6 Singular Fiberings

Throughout this section, Σ and $\mathcal{O}$ denote compact connected orbifolds, in the sense of Sect. 3.5.

Definition 3.12. A continuous surjection $p\colon \Sigma \to \mathcal{O}$ is called a *singular fibering* if there exists a commutative diagram

$$\begin{array}{ccc} \widetilde{\Sigma} & \xrightarrow{\widetilde{p}} & \widetilde{\mathcal{O}} \\ \downarrow{\sigma} & & \downarrow{\tau} \\ \Sigma & \xrightarrow{p} & \mathcal{O} \end{array}$$

in which

(i) $\widetilde{\Sigma}$ and $\widetilde{\mathcal{O}}$ are manifolds, and σ and τ are regular orbifold coverings with groups of covering transformations G and H respectively.
(ii) $\widetilde{p}$ is surjective and locally trivial.
(iii) The fibers of p and $\widetilde{p}$ are path-connected.

The class of singular fiberings includes many Seifert fiberings, for example all compact three-dimensional Seifert manifolds Σ except the lens spaces with one or two exceptional orbits (see for example [60]). For some of those lens spaces, $\mathcal{O}$ fails to have an orbifold covering by a manifold. On the other hand, it is a much larger class than Seifert fiberings, because no structure as a homogeneous space is required on the fiber.

For mappings there is a complete analogy with the bundle case, where now $\mathrm{Diff}_f(\Sigma)$ is by definition the quotient of the group of fiber-preserving G-equivariant diffeomorphisms $(\mathrm{Diff}_G)_f(\widetilde{\Sigma})$ by its normal subgroup G, and so on. A suborbifold W of Σ is called *vertical* if it is a union of fibers. In this case the inverse image $\widetilde{W}$ of W in $\widetilde{\Sigma}$ is a vertical submanifold, and we write $\mathrm{Emb}_f(W, \Sigma)$ for embeddings induced by elements of $(\mathrm{Emb}_G)_f(\widetilde{W}, \widetilde{\Sigma})$, $\mathrm{Emb}_v(W, \Sigma)$ for embeddings induced by elements of $(\mathrm{Emb}_G)_v(\widetilde{W}, \widetilde{\Sigma})$, and so on.

Following our usual notations, we put $\partial_v \Sigma = p^{-1}(\partial \mathcal{O})$, $\partial_v W = W \cap \partial_v \Sigma$, $\partial_h \Sigma = \overline{\partial \Sigma - \partial_v \Sigma}$, and $\partial_h W = \partial W \cap \partial_h \Sigma$.

Since $\mathcal{O}$ is compact, Lemma 3.8 shows that a (complete) Riemannian metric on $\widetilde{\mathcal{O}}$ can be chosen so that H acts as isometries, and moreover so that the metric on $\widetilde{\mathcal{O}}$ is a product near the boundary. Next we will sketch how to obtain a G-equivariant metric which is a product near $\partial_h \widetilde{\Sigma}$ and near $\partial_v \widetilde{\Sigma}$. If $\partial_h \widetilde{\Sigma}$ is empty, we simply apply Lemma 3.8. Assume that $\partial_h \widetilde{\Sigma}$ is nonempty. Construct a G-equivariant collar of $\partial_h \widetilde{\Sigma}$, and use it to obtain a G-equivariant metric such that the I-fibers of $\partial_h \widetilde{\Sigma} \times \mathrm{I}$ are vertical. If $\partial_v \widetilde{\Sigma}$ is also nonempty, put $Y = \partial_h \widetilde{\Sigma} \cap \partial_v \widetilde{\Sigma}$. We will follow the construction in the last paragraph of Sect. 2.6. Denote the collar of $\partial_h \widetilde{\Sigma}$ by $\partial_h \widetilde{\Sigma} \times [0, 2]_1$. Assume that the metric on $\partial_h \widetilde{\Sigma}$ was a product on a collar $Y \times [0, 2]_2$ of Y in $\partial_h \widetilde{\Sigma}$. Next, construct a G-equivariant collar $\partial_v \widetilde{\Sigma} \times [0, 2]_2$ of $\partial_v \widetilde{\Sigma}$ whose $[0, 2]_2$-fiber at each point of $Y \times [0, 2]_1$ agrees with the $[0, 2]_2$-fiber of the collar of Y in $\partial_h \widetilde{\Sigma} \times \{t\}$. Then, the product metric on $\partial_v \widetilde{\Sigma} \times [0, 2]_2$ agrees with the product metric of $\partial_h \widetilde{\Sigma} \times [0, 2]_1$ where they overlap, and the G-equivariant patching can be done to obtain a metric which is a product near $\partial_v \widetilde{\Sigma}$ without losing the property that it is a product near $\partial_h \widetilde{\Sigma}$. We will always assume that the metrics have been selected with these properties. By the first sentence of the next lemma, G preserves the vertical and horizontal parts of vectors.

Some basic observations about singular fiberings will be needed.

Lemma 3.12. *The action of G preserves the fibers of $\widetilde{p}$. Moreover:*

(i) *If $g \in G$, then there exists an element $h \in H$ such that $\widetilde{p}g = h\widetilde{p}$.*
(ii) *If $h \in H$, then there exists an element g of G such that $\widetilde{p}g = h\widetilde{p}$.*
(iii) *If $x \in \Sigma$, then $\tau^{-1}p(x) = \widetilde{p}\sigma^{-1}(x)$.*

Proof. Suppose that $\widetilde{p}(x) = \widetilde{p}(y)$. For $g \in G$, we have $\tau\widetilde{p}(g(x)) = p\sigma(g(x)) = p\sigma(x) = \tau\widetilde{p}(x) = \tau\widetilde{p}(y) = \tau\widetilde{p}(g(y))$. Since the fibers of $\widetilde{p}$ are path-connected, and the fibers of τ are discrete, this implies that $g(x)$ and $g(y)$ lie in the same fiber of $\widetilde{p}$. For (i), let $g \in G$. Since g preserves the fibers of $\widetilde{p}$, it induces a map h on $\widetilde{\mathcal{O}}$. Given $x \in \widetilde{\mathcal{O}}$, choose $y \in \widetilde{\Sigma}$ with $\widetilde{p}(y) = x$. Then $\tau h(x) = \tau\widetilde{p}(g(y)) = p\sigma(g(y)) = p\sigma(y) = \tau\widetilde{p}(y) = \tau(x)$ so $h \in H$.

To prove (ii), suppose h is any element of H. Let $\mathrm{sing}(\mathcal{O})$ denote the singular set of $\mathcal{O}$. Choose $a \in \widetilde{\mathcal{O}} - \tau^{-1}(\mathrm{sing}(\mathcal{O}))$, choose $s \in \widetilde{\Sigma}$ with $\widetilde{p}(s) = a$, and choose $s'' \in \widetilde{\Sigma}$ with $\widetilde{p}(s'') = h(a)$. Since $p\sigma(s) = \tau\widetilde{p}(s) = \tau\widetilde{p}(s'') = p\sigma(s'')$, $\sigma(s)$ and $\sigma(s'')$ must lie in the same fiber of p. Since the fiber is path-connected, there exists a path β in that fiber from $\sigma(s'')$ to $\sigma(s)$. Let $\widetilde{\beta}$ be its lift in $\widetilde{\Sigma}$ starting at s'' and let s' be the endpoint of this lift, so that $\sigma(s') = \sigma(s)$. Note that $\widetilde{p}(s') = \widetilde{p}(s'') = h(a)$ since $\widetilde{\beta}$ lies in a fiber of $\widetilde{p}$. Since $\sigma(s) = \sigma(s')$, there exists a covering transformation

$g \in G$ with $g(s) = s'$. To show that $\widetilde{p}g = h\widetilde{p}$, it is enough to verify that they agree on the dense set $\widetilde{p}^{-1}(\widetilde{\mathscr{O}} - \tau^{-1}(\mathrm{sing}(\mathscr{O})))$. Let $t \in \widetilde{p}^{-1}(\widetilde{\mathscr{O}} - \tau^{-1}(\mathrm{sing}(\mathscr{O})))$ and choose a path γ in $\widetilde{p}^{-1}(\widetilde{\mathscr{O}} - \tau^{-1}(\mathrm{sing}(\mathscr{O})))$ from s to t. Since $g \in G$, we have $p\sigma\gamma = p\sigma g\gamma$. Therefore $\tau\widetilde{p}\gamma = \tau\widetilde{p}g\gamma$, and so $\widetilde{p}g\gamma$ is the unique lift of $p\sigma\gamma$ starting at $\widetilde{p}g(s) = h(a)$. But this lift equals $h\widetilde{p}\gamma$, so $h\widetilde{p}(t) = \widetilde{p}g(t)$.

For (iii), fix $z_0 \in \sigma^{-1}(x)$ and let $y_0 = \widetilde{p}(z_0)$. Suppose $y \in \widetilde{p}\sigma^{-1}(x)$. Choose $z \in \sigma^{-1}(x)$ with $\widetilde{p}(z) = y$. Since σ is a regular covering, there exists $g \in G$ such that $g(z) = z_0$. By (i), g induces h on $\widetilde{\mathscr{O}}$, and $h(y) = h\widetilde{p}(z) = \widetilde{p}g(z) = \widetilde{p}(z_0) = y_0$. Therefore $\tau(y) = \tau(h(y)) = \tau(y_0) = \tau\widetilde{p}(z_0) = p\sigma(z_0) = p(x)$ so $y \in \tau^{-1}(p(x))$. For the opposite inclusion, suppose that $y \in \tau^{-1}p(x)$, so $\tau(y) = p(x) = \tau(y_0)$. Since σ is regular, there exists $h \in H$ such that $h(y_0) = y$. Let g be as in (ii). Then $y = h(y_0) = h\widetilde{p}(z_0) = \widetilde{p}g(z_0)$, and $\sigma(g(z_0)) = \sigma(z_0) = x$ so $y \in \widetilde{p}(\sigma^{-1}(x))$. □

One consequence of Lemma 3.12 is that there is a unique surjective homomorphism $\phi\colon G \to H$ with respect to which $\widetilde{p}$ is equivariant: $\widetilde{p}(gx) = \phi(g)(\widetilde{p}(x))$.

A second consequence of Lemma 3.12 is that provided that G acts as isometries, the aligned exponential Exp_a for the bundle $\widetilde{p}\colon \widetilde{\Sigma} \to \widetilde{\mathscr{O}}$ is G-equivariant. Consequently, the aligned tame exponential TExp_a takes G-equivariant vector fields on $\widetilde{\Sigma}$ to G-equivariant smooth maps of $\widetilde{\Sigma}$.

Theorem 3.8. *Let S be a closed subset of $\mathscr{O}$, and let $T = p^{-1}(S)$. Then* Diff $(\mathscr{O}$ rel $S)$ *admits local* $\mathrm{Diff}_f(\Sigma$ rel $T)$ *cross-sections.*

Proof. By Proposition 3.2, we only need a local $\mathrm{Diff}_f(\Sigma$ rel $T)$ cross-section at $1_{\mathscr{O}}$.

Applying Lemma 3.10 with $\mathscr{W} = \widetilde{\mathscr{O}}$ provides a neighborhood $\widetilde{U}$ of $1_{\widetilde{\mathscr{O}}}$ in $\mathrm{Diff}_H(\widetilde{\mathscr{O}}$ rel $\tau^{-1}(S))$ and $X\colon \widetilde{U} \to \mathscr{X}_H(\widetilde{\mathscr{O}}, T\widetilde{\mathscr{O}})$ such that $\mathrm{Exp}(X(j)(y)) = j(y)$ for all $y \in \widetilde{\mathscr{O}}$, and $X(j)(y) = Z(y)$ for all $y \in \tau^{-1}(S)$. Define $\widetilde{X}\colon \widetilde{U} \to \mathscr{X}(\widetilde{\Sigma}, T\widetilde{\Sigma})$ by taking horizontal lifts, that is,

$$\widetilde{X}(j)(x) = \left(\widetilde{p}|_{H_x}\right)_*^{-1}(X(j)(\widetilde{p}(x)))\ .$$

We claim that $\widetilde{X}(j)$ lies in $\mathscr{A}_G(\widetilde{\Sigma}, T\widetilde{\Sigma})$. To verify the boundary tangency conditions, we observe that $\widetilde{X}(j)$ must be tangent to the vertical boundary since it is a lift of a vector tangent to the boundary of $\widetilde{\mathscr{O}}$, and tangent to the horizontal boundary since it is horizontal. Since $\mathrm{Exp}(X(j)(y))$ is defined at all points of $\widetilde{\mathscr{O}}$, and $\widetilde{X}(j)$ is horizontal, each $\mathrm{Exp}_a(\widetilde{X}(j))(x))$ exists. To check equivariance, let $g \in G$. By Lemma 3.12, there exists $h \in H$ such that $\widetilde{p}g = h\widetilde{p}$. We then have

$$\begin{aligned}
\widetilde{X}(j)(g(x)) &= \left(\widetilde{p}|_{H_x}\right)_*^{-1}(X(j)(\widetilde{p}(g(x)))) \\
&= \left(\widetilde{p}|_{H_x}\right)_*^{-1}(X(j)(h\widetilde{p}(x))) \\
&= \left(\widetilde{p}|_{H_x}\right)_*^{-1}(h_*(X(j)(\widetilde{p}(x)))) \\
&= g_*\left(\widetilde{p}|_{H_x}\right)_*^{-1}(X(j)(h\widetilde{p}(x))) \\
&= g_*\widetilde{X}(j)(x)\ .
\end{aligned}$$

Note also that $\widetilde{X}(j)(x) = Z(x)$ for every $x \in \widetilde{p}^{-1}\tau^{-1}(S)$, since $X(\underset{\sim}{j})(y) = Z(y)$ for every $y \in \tau^{-1}(S)$. Using Lemma 3.9, we may pass to a smaller $\widetilde{U}$ if necessary to assume that $\mathrm{TExp}_a \circ \widetilde{X} : \widetilde{U} \to \mathrm{Diff}_G(\widetilde{\Sigma} \text{ rel } \widetilde{p}^{-1}\tau^{-1}(S))$.

Since $\tau : \widetilde{\mathcal{O}} \to \mathcal{O}$ is an orbifold covering map, there exists a neighborhood of $1_{\widetilde{\mathcal{O}}}$ in $\mathrm{Diff}_H(\widetilde{\mathcal{O}})$ such that no two elements in this neighborhood induce the same diffeomorphism on $\mathcal{O}$. Intersecting this neighborhood with $\widetilde{U}$, we may assume that $\widetilde{U}$ has the same property.

By definition, each $f \in \mathrm{Diff}(\mathcal{O})$ has lifts to elements of $\mathrm{Diff}_H(\widetilde{\mathcal{O}})$. If f lies in some sufficiently small neighborhood U of $1_{\mathcal{O}}$, then it has a lift in $\widetilde{U}$. This lift is unique, by our selection of $\widetilde{U}$, and we denote it by $\widetilde{f}$. Define $\chi : U \to \mathrm{Diff}(\Sigma \text{ rel } T)$ by letting $\chi(\underset{\sim}{f})$ be the diffeomorphism induced on Σ by $\mathrm{TExp}_a \circ \widetilde{X}(\widetilde{f})$. Let $y \in \mathcal{O}$, choose $\widetilde{y} \in \widetilde{\mathcal{O}}$ with $\tau(\widetilde{y}) = y$, and $\widetilde{x} \in \widetilde{\Sigma}$ with $\widetilde{p}(\widetilde{x}) = \widetilde{y}$. Then we have

$$
\begin{aligned}
(\chi(f) \cdot 1_{\mathcal{O}})(y) &= \overline{\chi(f)}(y) = \overline{\chi(f)}(\tau \circ \widetilde{p}(\widetilde{x})) = \overline{\chi(f)}(p \circ \sigma(\widetilde{x})) \\
&= p \circ \chi(f)(\sigma(\widetilde{x})) = p \circ \sigma \circ \mathrm{TExp}_a \circ \widetilde{X}(\widetilde{f})(\widetilde{x}) \\
&= \tau \circ \widetilde{p} \circ \mathrm{TExp}_a \circ \widetilde{X}(\widetilde{f})(\widetilde{x}) = \tau \circ \mathrm{Exp} \circ \widetilde{p}_* \circ \widetilde{X}(\widetilde{f})(\widetilde{x}) \\
&= \tau \circ \mathrm{Exp} \circ X(\widetilde{f})(\widetilde{y}) = \tau \circ \widetilde{f}(y) = f(y)
\end{aligned}
$$

as required. □

Applying Proposition 3.1, we have immediately

Theorem 3.9. *If S is a closed subset of $\mathcal{O}$ and $T = p^{-1}(S)$, then* $\mathrm{Diff}_f(\Sigma \text{ rel } T) \to \mathrm{Diff}(\mathcal{O} \text{ rel } S)$ *is locally trivial.*

We now examine Lemmas 3.4 and 3.5 in the singular fibered case. There is no difficulty in adapting Lemma 3.4 equivariantly:

Lemma 3.13 (Logarithm Lemma for singular fiberings). *Let W be a vertical suborbifold of Σ. Then there are an open neighborhood U of the inclusion $i_{\widetilde{W}}$ in $(\mathrm{Emb}_f)_G(\widetilde{W}, \widetilde{\Sigma})$ and a continuous map $X : U \to \mathcal{A}_G(\widetilde{W}, T\widetilde{\Sigma})$ such that for all $j \in U$, $\mathrm{Exp}_a(X(j)(x))$ is defined for all $x \in \widetilde{W}$ and $\mathrm{Exp}_a(X(j)(x)) = j(x)$ for all $x \in \widetilde{W}$. Also, $X(i_{\widetilde{W}}) = Z$.*

Lemma 3.14 (Extension Lemma for singular fiberings). *Let W be a vertical suborbifold of Σ, and T a closed fibered neighborhood in $\partial_v\Sigma$ of $T \cap \partial_v W$. Then there exists a continuous linear map $k : \mathcal{A}_G(\widetilde{W}, T\widetilde{\Sigma}) \to \mathcal{A}_G(\widetilde{\Sigma}, T\widetilde{\Sigma})$ such that $k(X)(x) = X(x)$ for all $X \in \mathcal{A}_G(\widetilde{W}, T\widetilde{\Sigma})$ and $x \in \widetilde{W}$. If $X(x) = Z(x)$ for all $x \in \widetilde{T} \cap \partial_v \widetilde{W}$, then $k(X)(x) = Z(x)$ for all $x \in \widetilde{T}$. Moreover, $k(\mathcal{V}_G(\widetilde{W}, T\widetilde{\Sigma})) \subset \mathcal{V}_G(\widetilde{\Sigma}, T\widetilde{\Sigma})$.*

Proof. We may assume that the metrics on $\widetilde{\Sigma}$ and $\widetilde{\mathcal{O}}$ are G- and H-equivariant; in particular, G takes horizontal subspaces of $T\widetilde{\Sigma}$ to horizontal subspaces. Notice that $\widetilde{p}_*$ carries G-invariant aligned vector fields to H-invariant vector fields; this uses Lemma 3.12(ii). It follows that the aligned exponential on $\widetilde{\Sigma}$ is G-equivariant.

For let $X \in \mathscr{A}_G(T\widetilde{\Sigma})$ and let $g \in G$. Let $x \in \widetilde{\Sigma}$ and let $\widetilde{F}_x$ be the fiber of $\widetilde{p}$ containing x. At x, $X(x) = X(x)_v + X(x)_h$. Since g is an isometry, $X(g(x))_v = g_*(X(x)_v)$ and $X(g(x))_h = g_*(X(x)_h)$. To find $\mathrm{Exp}_a(X(x))$, we first find $\mathrm{Exp}_v(X(x)_v)$, that is, exponentiate $X(x)_v$ using the metric induced on $\widetilde{F}_x$. This ends at a point $x' \in \widetilde{F}_x$. Since G acts as isometries, $\mathrm{Exp}_v(g_*X(x)_v) = g\,\mathrm{Exp}_v(X(x)_v) = g(x')$. Now, use Lemma 3.12 to obtain $\lambda \in H$ with $\lambda\widetilde{p} = \widetilde{p}g$. We have $\lambda_*\widetilde{p}_*(X(x)_h) = \widetilde{p}_*(g_*(X(x)_h)) = \widetilde{p}_*(X(g(x))_h)$. Since λ is an isometry, it carries the geodesic in $\mathscr{O}$ determined by $\widetilde{p}_*(X(x)_h)$ to the geodesic determined by $\widetilde{p}_*(X(g(x))_h)$. Therefore g carries the horizontal lift of $\widetilde{p}_*(X(x)_h)$ at x' to the horizontal lift of $\widetilde{p}_*(X(g(x))_h)$ at $g(x')$. So g carries $\mathrm{Exp}_a(X(x))$ to $\mathrm{Exp}_a(X(g(x)))$.

We can now proceed as in the proof of Lemma 3.5. Given a G-equivariant aligned section on W, extend the vertical part as in Lemma 3.2 and project the extension to the vertical subspace. This process is equivariant since we use a G-equivariant metric and G-equivariant functions to taper off the local extensions. For the horizontal part, project to $\widetilde{\mathscr{O}}$, extend H-equivariantly using Lemma 3.11, and lift. □

Theorem 3.10. *Let W be a vertical suborbifold of Σ. Let T be a closed fibered neighborhood in $\partial_v\Sigma$ of $T \cap \partial_v W$. Then*

(i) $\mathrm{Emb}_f(W, \Sigma \text{ rel } T)$ *admits local* $\mathrm{Diff}_f(\Sigma \text{ rel } T)$ *cross-sections.*
(ii) $\mathrm{Emb}_v(W, \Sigma \text{ rel } T)$ *admits local* $\mathrm{Diff}_v(\Sigma \text{ rel } T)$ *cross-sections.*

Proof. By Proposition 3.2, it suffices to find local cross-sections at the inclusion i_W.

By Lemma 3.13, there are an open neighborhood $\widetilde{U}$ of the inclusion $i_{\widetilde{W}}$ in $(\mathrm{Emb}_f)_G(\widetilde{W}, \widetilde{\Sigma})$ and a continuous map $X: \widetilde{U} \to \mathscr{A}_G(\widetilde{W}, T\widetilde{\Sigma})$ such that for all $j \in \widetilde{U}$, $\mathrm{Exp}_a(X(j)(x))$ is defined for all $x \in \widetilde{W}$ and $\mathrm{Exp}_a(X(j)(x)) = j(x)$ for all $x \in \widetilde{W}$. By Lemma 3.14, there exists a continuous linear map $k: \mathscr{A}_G(\widetilde{W}, T\widetilde{\Sigma}) \to \mathscr{A}_G(\widetilde{\Sigma}, T\widetilde{\Sigma})$ such that $k(X)(x) = X(x)$ for all $X \in \mathscr{A}_G(\widetilde{W}, T\widetilde{\Sigma})$ and $x \in \widetilde{W}$. Additionally, $k(X)(x) = Z(x)$ for all $x \in \widetilde{T}$, and $k(\mathscr{V}_G(\widetilde{W}, T\widetilde{\Sigma})) \subset \mathscr{V}_G(\widetilde{\Sigma}, T\widetilde{\Sigma})$.

Lemma 3.6 now gives a neighborhood $\widetilde{U}_1$ of Z in $\mathscr{A}_G(\widetilde{\Sigma}, T\widetilde{\Sigma})$ such $\mathrm{Exp}_a(X)$ is defined for all $X \in \widetilde{U}_1$, and TExp_a has image in $\mathrm{Diff}_f^K(E)$. Putting $U = X^{-1} \circ k^{-1}(\widetilde{U}_1)$, the composition $\mathrm{TExp}_a \circ k \circ X: \widetilde{U} \to \mathrm{Diff}_f^K(E)$ is the desired cross-section for (i). Since X carries $\mathrm{Emb}_v(W, \Sigma)$ into $\mathscr{V}_G(\widetilde{W}, T\widetilde{\Sigma})$, k carries $\mathscr{V}_G(\widetilde{W}, T\widetilde{\Sigma})$ into $\mathscr{V}_G(\widetilde{\Sigma}, T\widetilde{\Sigma})$, and TExp_a carries $\widetilde{U}_1 \cap \mathscr{V}_G(\widetilde{\Sigma}, T\widetilde{\Sigma})$ into $\mathrm{Diff}_v(\widetilde{\Sigma})$, this cross-section restricts on $\mathrm{Emb}_v(W, \Sigma \text{ rel } T)$ to a $\mathrm{Diff}_v^L(\Sigma \text{ rel } T)$ cross-section, giving (ii). □

As in Sect. 3.4, we have the following immediate corollaries.

Corollary 3.8. *Let W be a vertical suborbifold of Σ. Let T be a fibered neighborhood in $\partial_v\Sigma$ of $T \cap \partial_v W$. Then the following restrictions are locally trivial:*

(i) $\mathrm{Diff}_f(\Sigma \text{ rel } T) \to \mathrm{Emb}_f(W, \Sigma \text{ rel } T)$.
(ii) $\mathrm{Diff}_v(\Sigma \text{ rel } T) \to \mathrm{Emb}_v(W, \Sigma \text{ rel } T)$.

Corollary 3.9. *Let V and W be vertical suborbifolds of Σ, with $W \subseteq V$. Let T be a closed fibered neighborhood in $\partial_v\Sigma$ of $T \cap \partial_v W$. Then the following restrictions are locally trivial:*

(i) $\mathrm{Emb}_f(V, \Sigma \text{ rel } T) \to \mathrm{Emb}_f(W, \Sigma \text{ rel } T)$.
(ii) $\mathrm{Emb}_v(V, \Sigma \text{ rel } T) \to \mathrm{Emb}_v(W, \Sigma \text{ rel } W)$.

Theorem 3.11. *Let W be a vertical suborbifold of Σ. Let T be a closed fibered neighborhood in $\partial_v\Sigma$ of $T \cap \partial_v W$, and let $S = p(T)$. Then all four maps in the following square are locally trivial:*

$$\begin{array}{ccc}
\mathrm{Diff}_f(\Sigma \text{ rel } T) & \longrightarrow & \mathrm{Emb}_f(W, \Sigma \text{ rel } T) \\
\downarrow & & \downarrow \\
\mathrm{Diff}(\mathcal{O} \text{ rel } S) & \longrightarrow & \mathrm{Emb}(p(W), \mathcal{O} \text{ rel } S)\,.
\end{array}$$

3.7 Spaces of Fibered Structures

In this section, we examine spaces of fibered structures.

Definition 3.13. Let $p\colon \Sigma \to \mathcal{O}$ be a singular fibering. The *space of fibered structures* isomorphic to p, (also called the *space of singular fiberings* isomorphic to p) is the space of cosets $\mathrm{Diff}(\Sigma)/\mathrm{Diff}_f(\Sigma)$.

Our proof of the next theorem requires an additional condition, although we do not know that it is necessary:

Definition 3.14. A singular fibering $p\colon \Sigma \to \mathcal{O}$ is called *very good* if $\widetilde{\Sigma}$ may be chosen to be compact.

The main result of this section is the following fibration theorem.

Theorem 3.12. *Let $p\colon \Sigma \to \mathcal{O}$ be a very good singular fibering. Then the space of fibered structures isomorphic to p is a Fréchet manifold locally modeled on the quotient $\mathscr{X}_G(\widetilde{\Sigma}, T\widetilde{\Sigma})/\mathscr{A}_G(\widetilde{\Sigma}, T\widetilde{\Sigma})$. The quotient map* $\mathrm{Diff}(\Sigma) \to \mathrm{Diff}(\Sigma)/\mathrm{Diff}_f(\Sigma)$ *is a locally trivial fibering.*

Here is the basic idea of the proof. Roughly speaking, finding a local $\mathrm{Diff}(\Sigma)$ cross-section for $\mathrm{Diff}(\Sigma)/\mathrm{Diff}_f(\Sigma)$ boils down to the problem of taking an $h \in \mathrm{Diff}(\Sigma)$ that carries fibers of Σ to fibers that are nearly vertical, and finding, for each fiber F of Σ, a "nearest" vertical fiber to $h(F)$. It is not obvious that such a choice is uniquely determined, but there is a way to make one when h is sufficiently close to a fiber-preserving diffeomorphism. For then each $p(h(F))$ lies a very small open ball set in B, and $p(h(F))$ has a unique *center of mass* $c_{p(h(F))}$. The natural choice for the nearest fiber to $h(F)$ is $p^{-1}(c_{p(h(F))})$.

Before beginning the proof, we must clarify the idea of center of mass in this context. A useful reference for this is Karcher [40], which we will follow here.

Let A be a measure space of volume 1 and let B be an open ball in a compact Riemannian manifold M. By making its radius small enough, we may ensure that the closure $\overline{B}$ is a geodesically convex ball (that is, any two points in $\overline{B}$ are connected by a unique geodesic that lies in $\overline{B}$). Let $f: A \to M$ be a measurable map such that $f(A) \subset B$. Define $P_f: \overline{B} \to \mathbb{R}$ by

$$P_f(m) = \frac{1}{2} \int_A d(m, f(a))^2 \, dA .$$

Various estimates on the gradient of P_f, detailed in [40], show that P_f is a convex function that has a unique minimum in B, and this minimum is defined to be the center of mass C_f of f. From its definition, C_f is independent of the choice of B, although it is the existence of such a B that serves to ensure that it is uniquely defined.

Proof (of Theorem 3.12). Consider first the case of an ordinary bundle $p: E \to B$ with E compact. For each $x \in E$, the fiber containing x will denoted by F_x. For each coset $h \operatorname{Diff}_f(E)$, the set of images $\{h(F_x)\}$ is independent of the coset representative, and we will refer to these submanifolds as "image fibers", reserving "fibers" for the original fibers for p. When the coset $h \operatorname{Diff}_f(E)$ is clear from the context, the image fiber containing x will be denoted by F'_x.

Write n for the dimension of E and k for the dimension of B. The tangent bundle of E has an associated bundle $G_k(TE)$ whose fiber is the Grassmannian of k-planes in $\mathbb{R}^n$, and selecting the horizontal k-plane at each point defines a section $s_0: E \to G_k(TEy)$. The normal subspaces for the image fibering of $h \operatorname{Diff}_f(E)$ determine another section $s: E \to G_k(TE)$, defining a function $\operatorname{Diff}(E)/\operatorname{Diff}_f(E) \to C^\infty(E, G_k(TE))$. This function is injective, since distinct fiberings must have different normal spaces at some points, so imbeds $\operatorname{Diff}(E)/\operatorname{Diff}_f(E)$ into the Fréchet space of sections from M into $G_k(TEy)$. In particular, we can speak of image fiberings as being C^∞-close to vertical, meaning that the section s is C^∞-close to s_0.

We will first produce local $\operatorname{Diff}(E)$ cross-sections, then examine the Fréchet structure on $\operatorname{Diff}(E)/\operatorname{Diff}_f(E)$. Since $\operatorname{Diff}(E)$ acts transitively on $\operatorname{Diff}(E)/\operatorname{Diff}_f(E)$, it is enough to produce a local cross-section at the identity coset $1_E \operatorname{Diff}_f(E)$.

For $\varepsilon > 0$, denote by $H_\varepsilon(F_x)$ the space of horizontal vectors in $TE|_{F_x}$ of length less than ε that are carried into E by the aligned exponential Exp_a. By compactness, there exists an $\varepsilon_0 > 0$ such that for every $x \in E$, Exp_a carries $H_{\varepsilon_0}(F_x)$ diffeomorphically onto a tubular neighborhood $N_{\varepsilon_0}(F_x)$ of F_x in E. We may also choose ε_0 so that each ball in B of radius at most ε_0 has convex closure.

By compactness of E, there exists a neighborhood U of $1_E \operatorname{Diff}_f(E)$ in the coset space $\operatorname{Diff}(E)/\operatorname{Diff}_f(E)$ such that for each $h \operatorname{Diff}_f(E) \in U$, the image fibering of $h \operatorname{Diff}_f(E)$ has the following property: For each $y \in E$, there exists a fiber F_x such

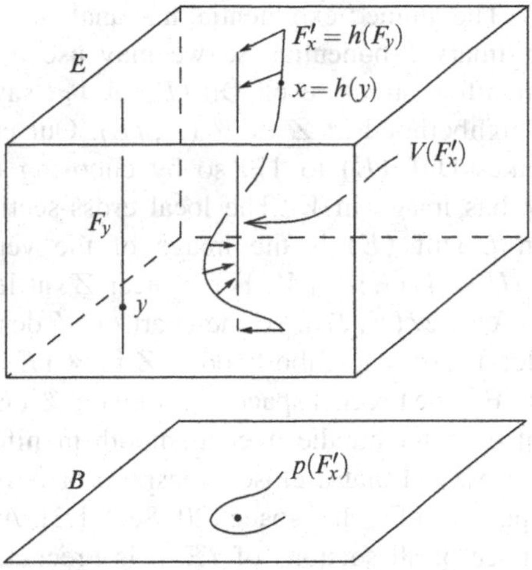

Fig. 3.2 Canonical straightening of a nearly vertical fiber. The *dot* in B is the center of mass of the projection $p(F'_x)$ of the image fiber F'_x. The inverse image of the center of mass is the straightened fiber $V(F'_x)$, and some of the horizontal vector field X is shown

that $F'_y \subset N_{\varepsilon_0}(F_x)$, and moreover if F_x is any such fiber, then F'_y meets each normal fiber of $N_{\varepsilon_0}(F_x)$ transversely in a single point.

Now we will set up the center-of-mass construction, illustrated in Fig. 3.2. Fix a coset $h\,\mathrm{Diff}_f(E)$ and an image fiber F'_x, where $x = h(y)$ for some y. Let dF'_x be the volume form on F'_x obtained from restriction of the Riemannian metric on TE to TF'_x, and define a measure $\mu_{F'_x}$ on F'_x of volume 1 by $\mu_{F'_x}(U) = \mathrm{Vol}(U)/\,\mathrm{Vol}(F'_x)$.

Assume now that $h\ \mathrm{Diff}_f(E)$ is close enough to vertical that for each image fiber F'_x, $p(F'_x)$ lies in some ε_0-ball. The center of mass of $(F'_x, m_{F'_x})$ is then defined, and we denote its inverse image, a fiber of p, by $V(F'_x)$.

For each $z \in E$, let $n(z)$ be the point of $V(F'_z)$ such that the normal fiber of $N_{\varepsilon_0}(V(F'_z))$ at $n(z)$ contains z. There is a unique horizontal vector $X(z) \in T_zE$ such that $\mathrm{Exp}_a(X(z)) = n(z)$. To see that the resulting horizontal vector field X is smooth, we first observe that changes of z along the image fiber simply correspond to changes of $n(z)$ along the fiber $V(F'_z)$. As z moves from image fiber to image fiber, the projected images in B of the image fibers are the images of the original fibers of E under the smooth map $p \circ h$. The corresponding centers of mass change smoothly, and the remainder of the construction presents no danger of loss of smoothness. Precomposing h by a fiber-preserving diffeomorphism does not change the image fibers, so $X(h\,\mathrm{Diff}_f(E))$ is well-defined. If h is fiber-preserving, then $V(F'_z) = F'_z$, $n(z) = z$, and $X(h) = Z$.

For each image fibering $h\,\mathrm{Diff}_f(E)$ in some C^∞-neighborhood $U\,\mathrm{Diff}_f(E)$ of $1_E\,\mathrm{Diff}_f(E)$, we have defined a horizontal vector field $X(h\,\mathrm{Diff}_f(E))$, for which applying the tame aligned exponential defines a smooth map $g_{h\,\mathrm{Diff}_f(E)}$ that moves each image fiber onto a vertical fiber. Since the coset $1_E\,\mathrm{Diff}_f(E)$ determines the zero vector field, $g_{\mathrm{Diff}_f(E)} = 1_E$. So by reducing the size of U, if necessary, each $g_{h\,\mathrm{Diff}_f(E)}$ will be a diffeomorphism. A local $\mathrm{Diff}(E)$ cross-section $\chi\colon U\,\mathrm{Diff}_f(E) \to \mathrm{Diff}(E)$ is then defined by sending $h\,\mathrm{Diff}_f(E)$ to $g^{-1}_{h\,\mathrm{Diff}_f(E)}$.

The aligned exponential has analogous local diffeomorphism properties to the ordinary exponential, so we may use it to define a local chart for the Fréchet manifold structure on Diff(E) at 1_E, say $\mathrm{TExp}_a\colon V \to \mathrm{Diff}(E)$, where V is a neighborhood of Z in $\mathscr{X}(E,TE)$. Our cross-section $\chi\colon U\,\mathrm{Diff}_f(E) \to \mathrm{Diff}(E)$ takes $\mathrm{Diff}_f(E)$ to 1_E, so by choosing U small enough, we may assume that χ has image in V. The local cross-section shows that every fibering contained in $U\,\mathrm{Diff}_f(E)$ is the image of the vertical fibering under a diffeomorphism $\chi(U\,\mathrm{Diff}_f(E))$ in V. For X near Z, at least, $\mathrm{TExp}_a(X) \in \mathrm{Diff}_f(E)$ if and only if $X \in \mathscr{A}(E,TE)$, so the chart on V descends to a chart for $\mathrm{Diff}(E)/\mathrm{Diff}_f(E)$, defined on a neighborhood of Z in $\mathscr{X}(E,TEy)/\mathscr{A}(E,TE)$.

For the Fréchet space structure on $\mathscr{X}(E,TE)/\mathscr{A}(E,TE)$, recall that the sections of a vector bundle over a smooth manifold form a Fréchet space [20, Example 1.1.5], and that a closed subspace or quotient by a closed subspace of a Fréchet space is a Fréchet space [20, Sect. 1.2]. As $\mathscr{X}(E,TE)$ is a closed subspace of the space of all sections of TE, it is Fréchet. Since $\mathscr{A}(E,TE)$ is a closed subspace, $\mathscr{X}(E,TE)/\mathscr{A}(E,TE)$ is Fréchet as well.

In the case of a very good singular fibering $p\colon \Sigma \to \mathscr{O}$, we carry out the previous construction working equivariantly in the bundle $\widetilde{\Sigma} \to \widetilde{\mathscr{O}}$, which may be chosen with $\widetilde{\Sigma}$ compact. Since we are using a G-equivariant Riemannian metric on $\widetilde{\Sigma}$ and an H-equivariant one on $\widetilde{\mathscr{O}}$, and $\widetilde{p}$ is equivariant, all parts of the construction proceed equivariantly. Because $\widehat{\Sigma}$ is compact, the image fibers of a G-equivariant diffeomorphism of $\widetilde{\Sigma}$ will project under $\widetilde{p}$ to compact sets, which are small when the fibers are nearly vertical, and consequently the centers of mass will still be well-defined. □

3.8 Restricting to the Boundary or the Basepoint

Our restriction theorems deal with the case when the suborbifold is properly embedded. By a simple doubling trick, we can also extend to restriction to suborbifolds of the boundary.

Proposition 3.4. *Let $\Sigma \to \mathscr{O}$ be a singular fibering. Let S be a suborbifold of $\partial\,\mathscr{O}$, and let $T = p^{-1}(S)$. Then*

(a) $\mathrm{Emb}(S, \partial\,\mathscr{O})$ *admits local* $\mathrm{Diff}(\mathscr{O})$ *cross-sections.*
(b) $\mathrm{Emb}_f(T, \partial_v \Sigma)$ *admits local* $\mathrm{Diff}_f(\Sigma)$ *cross-sections.*

Proof. We first show that $\mathrm{Diff}(\partial\,\mathscr{O})$ admits local $\mathrm{Diff}(\mathscr{O})$ cross-sections. Let Δ be the double of $\mathscr{O}$ along $\partial\,\mathscr{O}$, and regard $\mathscr{O}$ as a suborbifold of Δ by identifying it with one of the two copies of $\mathscr{O}$ in Δ. By Theorem 3.7, $\mathrm{Emb}(\partial\,\mathscr{O}, \Delta)$ admits local $\mathrm{Diff}(\Delta)$ cross-sections. We may regard $\mathrm{Diff}(\partial\,\mathscr{O})$ as a subspace of $\mathrm{Emb}(\partial\,\mathscr{O}, \Delta)$. Suppose that $\chi\colon U \to \mathrm{Diff}(\Delta)$ is a local cross-section at a point in $\mathrm{Emb}(\partial\,\mathscr{O}, \Delta)$ that lies in $\mathrm{Diff}(\partial\mathscr{O})$. Elements of $\mathrm{Diff}(\Delta, \partial\,\mathscr{O})$ that interchange the sides of $\mathscr{O}$ are far from elements that preserve the sides, so by making U smaller if necessary, we may assume that all elements $f \in U$ such that $\chi(f)$ lies in $\mathrm{Diff}(\Delta, \mathcal{O})$ either preserve the sides of $\mathcal{O}$ or interchange them. In the latter case, we postcompose χ with the diffeomorphism of Δ that interchanges the two copies of $\mathcal{O}$, to assume that all such elements preserve the sides. Then, sending g to $\chi(g)|_{\mathcal{O}}$ defines a local $\mathrm{Diff}(\mathcal{O})$ cross-section on $U \cap \mathrm{Diff}(\partial \mathcal{O})$.

By Proposition 3.2, for (a) it suffices to produce local cross-sections at the inclusion i_S. By Theorem 3.7, there is a local $\mathrm{Diff}(\partial \mathcal{O})$ cross-section χ_1 for $\mathrm{Emb}(S, \partial \mathcal{O})$ at i_S. Let χ_2 be a local $\mathrm{Diff}(\mathcal{O})$ cross-section for $\mathrm{Diff}(\partial \mathcal{O})$ at $\chi_1(i_S)$. On a neighborhood U of i_S in $\mathrm{Emb}(S, \partial \mathcal{O})$ small enough so that $\chi_2\chi_1$ is defined, the composition is the desired $\mathrm{Diff}(\mathcal{O})$ cross-section. For if $j \in U$, then $\chi_2(\chi_1(j)) \circ i_S = \chi_2(\chi_1(j)) \circ \chi_1(i_S) \circ i_S = \chi_1(j) \circ i_S = j$.

The proof of (b) is similar. Double Σ along $\partial_v \Sigma$ and apply Theorem 3.10, obtaining local $\mathrm{Diff}_f(\Sigma)$ cross-sections for $\mathrm{Diff}_f(\partial_v \Sigma)$. Apply it again to produce local $\mathrm{Diff}_f(\partial_v \Sigma)$ cross-sections for $\mathrm{Emb}_f(T, \partial_v \Sigma)$. Their composition, where defined, is a local $\mathrm{Diff}_f(\Sigma)$ cross-section for $\mathrm{Emb}_f(T, \partial_v \Sigma)$. □

An immediate consequence is

Corollary 3.10. *For a singular fibering $\Sigma \to \mathcal{O}$, let S be a suborbifold of $\partial \mathcal{O}$, and let $T = p^{-1}(S)$. Then* $\mathrm{Diff}(\mathcal{O}) \to \mathrm{Emb}(S, \partial \mathcal{O})$ *and* $\mathrm{Diff}_f(\Sigma) \to \mathrm{Emb}_f(T, \partial_v \Sigma)$ *are locally trivial. In particular,* $\mathrm{Diff}(\mathcal{O}) \to \mathrm{Diff}(\partial \mathcal{O})$ *and* $\mathrm{Diff}_f(\Sigma) \to \mathrm{Diff}_f(\partial_v \Sigma)$ *are locally trivial.*

Another consequence is

Corollary 3.11. *For a suborbifold $\mathcal{W}$ of $\mathcal{O}$, the restriction* $\mathrm{Emb}(\mathcal{W}, \mathcal{O}) \to \mathrm{Emb}(\mathcal{W} \cap \partial \mathcal{O}, \partial \mathcal{O})$ *is locally trivial.*

Proof. By Theorem 3.7, $\mathrm{Emb}(\mathcal{W} \cap \partial \mathcal{O}, \partial \mathcal{O})$ admits local $\mathrm{Diff}(\partial \mathcal{O})$ cross-sections, and by Proposition 3.4, $\mathrm{Diff}(\partial \mathcal{O})$ admits local $\mathrm{Diff}(\mathcal{O})$ cross-sections. Composing them gives local $\mathrm{Diff}(\mathcal{O})$ cross-sections for $\mathrm{Emb}(\mathcal{W} \cap \partial \mathcal{O}, \partial \mathcal{O})$. □

Corollary 3.12. *For a vertical suborbifold W of Σ, the restriction* $\mathrm{Emb}_f(W, \Sigma) \to \mathrm{Emb}_f(W \cap \partial_v \Sigma, \partial_v \Sigma)$ *is locally trivial.*

Proof. The restriction map is $\mathrm{Diff}_f(\Sigma)$-equivariant, and Proposition 3.4(b) shows that $\mathrm{Emb}_f(W \cap \partial_v \Sigma, \partial_v \Sigma)$ admits local $\mathrm{Diff}_f(\Sigma)$ cross-sections. □

Some applications of the fibration $\mathrm{Diff}(M) \to \mathrm{Emb}(V, M)$ concern the case when the submanifold is a single point. Since in the fibered case a single point is not usually a vertical submanifold, this case is not directly covered by our previous theorems. The next proposition allows nonvertical suborbifolds that are contained in a single fiber, so applies when the submanifold is a single point. To set notation, let $p \colon \Sigma \to \mathcal{O}$ be a singular fibering. Let P be a (properly-imbedded) suborbifold of Σ which is contained in a single fiber F. Let T be a fibered closed subset of $\partial_v \Sigma$ which does not meet F. By $\mathrm{Emb}_t(P, \Sigma - T)$ we denote the orbifold embeddings whose image is contained in a single fiber of $\Sigma - T$, and which extend to elements of $\mathrm{Diff}_f(\Sigma \text{ rel } T)$.

Proposition 3.5. *Let P be a suborbifold of Σ which is contained in a single fiber F. Let T be a fibered closed subset of $\partial_v \Sigma$, which does not meet F. Then* $\mathrm{Emb}_t(P, \Sigma - T)$ *admits local* $\mathrm{Diff}_f(\Sigma \text{ rel } T)$ *cross-sections.*

Proof. Let $S = p(T)$. Notice that $p(P)$ is a point and is a properly embedded suborbifold of $\mathcal{O}$, with orbifold structure determined by the local group at $p(P)$. Each embedding $i \in \mathrm{Emb}_t(P, \Sigma)$ induces an orbifold embedding $\overline{\imath}\colon p(P) \to \mathcal{O} - S$.

By Proposition 3.2, it suffices to produce a local cross-section at the inclusion i_P. By Theorem 3.7, $\mathrm{Emb}(p(P), \mathcal{O} - S)$ has local $\mathrm{Diff}(\mathcal{O} \text{ rel } S)$ cross-sections, and by Proposition 3.8, $\mathrm{Diff}(\mathcal{O} \text{ rel } S)$ has local $\mathrm{Diff}_f(\Sigma \text{ rel } T)$ cross-sections. A suitable composition of these gives a local $\mathrm{Diff}_f(\Sigma \text{ rel } T)$ cross-section χ_1 for $\mathrm{Emb}(p(P), \mathcal{O} - S)$ at $\overline{\imath_P}$. As remarked in Sect. 3.1, we may assume that $\chi_1(\overline{\imath_P})$ is the identity diffeomorphism of Σ. By Corollary 3.6, there exists a local $\mathrm{Diff}(F)$ cross-section χ_2 for $\mathrm{Emb}(P, F)$ at i_P, and we may assume that $\chi_2(i_P)$ is the identity diffeomorphism of F. Let χ_3 be a local $\mathrm{Diff}_f(\Sigma \text{ rel } T)$ cross-section for $\mathrm{Emb}_f(F, \Sigma - T)$ at i_F given by Corollary 3.8. Regarding $\mathrm{Diff}(F)$ as a subspace of $\mathrm{Emb}_f(F, \Sigma - T)$, we may assume that the composition $\chi_3 \chi_2$ is defined. On a sufficiently small neighborhood of i_P in $\mathrm{Emb}_t(P, \Sigma - T)$ define $\chi(j) \in \mathrm{Diff}_f(\Sigma \text{ rel } T)$ by

$$\chi(j) = \chi_1(p(j))\,(\chi_3 \chi_2)(\chi_1(p(j))^{-1} \circ j)\,.$$

Then for $x \in P$ we have

$$\begin{aligned} \chi(j) \circ i_P(x) &= \chi_1(p(j))\,(\chi_3\chi_2)(\chi_1(p(j))^{-1} \circ j) \circ i_P(x) \\ &= \chi_1(p(j))\,\chi_1(p(j))^{-1} \circ j(x) \\ &= j(x) \end{aligned}$$

□

This yields immediately

Corollary 3.13. *If W is a vertical suborbifold of Σ and P is a suborbifold of W contained in a fiber of W, then the maps* $\mathrm{Diff}_f(\Sigma \text{ rel } T) \to \mathrm{Emb}_t(P, \Sigma - T)$ *and* $\mathrm{Emb}_f(W, \Sigma \text{ rel } T) \to \mathrm{Emb}_t(P, \Sigma - T)$ *induced by restriction are locally trivial.*

3.9 The Space of Seifert Fiberings of a Haken Three-Manifold

Let $p\colon \Sigma \to \mathcal{O}$ be a Seifert fibering of a Haken manifold Σ. As noted in Sect. 3.6, p is a singular fibering. Denote by $\mathrm{diff}_f(\Sigma)$ the connected component of the identity in $\mathrm{Diff}_f(\Sigma)$, and similarly for other spaces of diffeomorphisms and embeddings. The main result of this section is the following.

Theorem 3.13. *Let Σ be a Haken Seifert-fibered three-manifold. Then the inclusion* $\mathrm{diff}_f(\Sigma) \to \mathrm{diff}(\Sigma)$ *is a homotopy equivalence.*

Before proving Theorem 3.13, we will derive some consequences. Each element of $\mathrm{Diff}(\Sigma)$ carries the given fibering to an isomorphic fibering, and $\mathrm{Diff}_f(\Sigma)$ is precisely the stabilizer of the given fibering under this action. Following Definition 3.13, we define the *space of Seifert fiberings* isomorphic to the given fibering to be the space of cosets $\mathrm{Diff}(\Sigma)/\mathrm{Diff}_f(\Sigma)$. Since Σ is not a lens space with one or two exceptional fibers, Σ is a singular fibering. Moreover, every Seifert fibering other than the exceptional lens space ones is finitely covered by an S^1-bundle (because apart from these cases, the quotient orbifold has a finite orbifold covering by a manifold), so is a very good singular fibering. So Theorem 3.12 ensures that the space of Seifert fiberings isomorphic to the given one is a separable Fréchet manifold, and the map

$$\mathrm{Diff}(\Sigma) \to \mathrm{Diff}(\Sigma)/\mathrm{Diff}_f(\Sigma)$$

is a fibration. Note that since $\mathrm{Diff}(\Sigma)/\mathrm{Diff}_f(\Sigma)$ is a Fréchet manifold, each connected component is a path component, and since $\mathrm{Diff}(\Sigma)$ acts transitively on $\mathrm{Diff}(\Sigma)/\mathrm{Diff}_f(\Sigma)$, any two components are homeomorphic.

Theorem 3.14. *Let Σ be a Seifert-fibered Haken three-manifold. Then each component of the space of Seifert fiberings of Σ is contractible.*

Proof. As sketched on p. 85 of [71], two fiber-preserving diffeomorphisms of Σ that are isotopic are isotopic through fiber-preserving diffeomorphisms. That is, $\mathrm{Diff}_f(\Sigma) \cap \mathrm{diff}(\Sigma) = \mathrm{diff}_f(\Sigma)$. Therefore the connected component of the identity in $\mathrm{Diff}(\Sigma)/\mathrm{Diff}_f(\Sigma)$ is $\mathrm{diff}(\Sigma)/(\mathrm{Diff}_f(\Sigma) \cap \mathrm{diff}(\Sigma)) = \mathrm{diff}(\Sigma)/\mathrm{diff}_f(\Sigma)$. Using Theorem 3.13, the latter is contractible. □

Theorem 3.14 shows that the space of Seifert fiberings of Σ is contractible when $\mathrm{Diff}_f(\Sigma) \to \mathrm{Diff}(\Sigma)$ is surjective, that is, when every self-diffeomorphism of Σ is isotopic to a fiber-preserving diffeomorphism. Almost all Haken Seifert-fibered three-manifolds have this property. The closed case is due to Waldhausen [69] (see also [49, Theorem 8.1.7]), who showed that (among Haken manifolds) it fails only for the three-torus, the double of the I-bundle over the Klein bottle with orientable total space, and the *Hantsche–Wendt* manifold, which is the manifold given by the Seifert invariants $\{-1; (n_2, 1); (2, 1), (2, 1)\}$ (see [49, pp. 133, 138], [12, pp. 478–481], [21, 69]). Topologically, the Hantsche–Wendt manifold is obtained by taking two copies of the I-bundle over the Klein bottle with orientable total space, one with the meridional fibering (the nonsingular fibering as an S^1-bundle over the Möbius band) and one with the longitudinal fibering (over the disk with two exceptional orbits of type $(2, 1)$) and gluing them together preserving the fibers on the boundary. It admits a diffeomorphism interchanging the two halves, which is not isotopic to a fiber-preserving diffeomorphism. For the bounded case, only S^1-bundles over the disk, annulus or Möbius band fail to have the property. This appears as Theorem VI.18 of Jaco [37]. We conclude:

Theorem 3.15. *Let Σ be a Seifert-fibered Haken three-manifold other than the Hantsche–Wendt manifold, the three-torus, the double of the* I*-bundle over the Klein* *bottle with orientable total space, or an S^1-bundle over the disk, annulus or Möbius band. Then* $\mathrm{Diff}_f(\Sigma) \to \mathrm{Diff}(\Sigma)$ *is a homotopy equivalence, that is, the space of Seifert fiberings of* Σ *is contractible.*

The remainder of this section will constitute the proof of Theorem 3.13.

Lemma 3.15. *Let* Σ *be a Seifert-fibered Haken three-manifold, and let* C *be a fiber of* Σ*. Then each component of* $\mathrm{Diff}_v(\Sigma \text{ rel } C)$ *is contractible.*

Proof. Since Σ is Haken, the base orbifold of $\Sigma - C$ has negative Euler characteristic and is not closed. It follows (see [60]) that $\Sigma - C$ admits an $\mathbb{H}^2 \times \mathbb{R}$ geometry. Thus there is an action of $\pi_1(\Sigma - C)$ on $\mathbb{H}^2 \times \mathbb{R}$ such that every element preserves the $\mathbb{R}$-fibers and acts as an isometry in the $\mathbb{H}^2$-coordinate.

It suffices to show that $\mathrm{diff}_v(\Sigma \text{ rel } C)$ is contractible. Let N be a fibered solid torus neighborhood of C in Σ. It is not difficult to see (as in the argument below) that $\mathrm{diff}_v(\Sigma \text{ rel } C)$ deformation retracts to $\mathrm{diff}_v(\Sigma \text{ rel } N)$, which can be identified with $\mathrm{diff}_v(\Sigma - C \text{ rel } N - C)$, so it suffices to show that the latter is contractible. For $f \in \mathrm{diff}_v(\Sigma - C \text{ rel } N - C)$, let F be a lift of f to $\mathbb{H}^2 \times \mathbb{R}$ that has the form $F(x,s) = (x, s + F_2(x,s))$, where $F_2(x,s) \in \mathbb{R}$.

Since f is vertically isotopic to the identity relative to $N - C$, we may moreover choose F so that $F_2(x,s) = 0$ if (x,s) projects to $N - C$. To see this, we choose the lift F to fix a point in the inverse image W of $N - C$. Since f is homotopic to the identity relative to $N - C$, F is equivariantly homotopic to a covering translation relative to W. That covering translation fixes the point in W, and therefore must be the identity. Thus F fixes W and commutes with every covering translation.

Define K_t by $K_t(x,s) = (x, s + (1-t)F_2(x,s))$. Since $K_0 = F$ and K_1 is the identity, and each K_t is the identity on the inverse image of $N - C$, this will define a contraction of $\mathrm{Diff}_v(\Sigma - C \text{ rel } N - C)$ once we have shown that each K_t is equivariant. Let $\gamma \in \pi_1(\Sigma - C)$. From [60], $\mathrm{Isom}(\mathbb{H}^2 \times \mathbb{R}) = \mathrm{Isom}(\mathbb{H}^2) \times \mathrm{Isom}(\mathbb{R})$, so we can write $\gamma(x,s) = (\gamma_1(x), \varepsilon_\gamma s + \gamma_2)$, where $\varepsilon_\gamma = \pm 1$ and $\gamma_2 \in \mathbb{R}$. Since $F\gamma = \gamma F$, a straightforward calculation shows that

$$F_2(\gamma_1(x), \varepsilon_\gamma s + \gamma_2) = \varepsilon_\gamma F_2(x,s) .$$

Now we calculate

$$\begin{aligned} K_t\gamma(x,s) &= K_t(\gamma_1(x), \varepsilon_\gamma s + \gamma_2) \\ &= (\gamma_1(x), \varepsilon_\gamma s + \gamma_2 + (1-t)F_2(\gamma_1(x), \varepsilon_\gamma s + \gamma_2)) \\ &= (\gamma_1(x), \varepsilon_\gamma s + \gamma_2 + (1-t)\varepsilon_\gamma F_2(x,s)) \\ &= (\gamma_1(x), \varepsilon_\gamma (s + (1-t)F_2(x,s)) + \gamma_2) \\ &= \gamma(x, s + (1-t)F_2(x,s)) \\ &= \gamma K_t(x,s) \end{aligned}$$

showing that K_t is equivariant. □

Proof (of Theorem 3.13). We first examine $\mathrm{diff}_v(\Sigma)$. Choose a regular fiber C and consider the restriction $\mathrm{diff}_v(\Sigma) \to \mathrm{emb}_v(C, \Sigma) \cong \mathrm{diff}(C) \cong \mathrm{diff}(S^1) \simeq \mathrm{SO}(2)$. By Corollary 3.8(ii), this is a fibration. By Lemma 3.15, each component of the fiber $\mathrm{Diff}_v(\Sigma \text{ rel } C) \cap \mathrm{diff}_v(\Sigma)$ is contractible. It follows by the exact sequence for this fibration that $\pi_q(\mathrm{diff}_v(\Sigma)) \cong \pi_q(\mathrm{SO}(2)) = 0$ for $q \geq 2$, and for $q = 1$ we have an exact sequence

$$0 \longrightarrow \pi_1(\mathrm{diff}_v(\Sigma)) \longrightarrow \pi_1(\mathrm{diff}(C)) \longrightarrow \pi_0(\mathrm{Diff}_v(\Sigma \text{ rel } C) \cap \mathrm{diff}_v(\Sigma)) \longrightarrow 0\,.$$

We will first show that exactly one of the following holds.

(a) C is central and $\pi_1(\mathrm{diff}_v(\Sigma)) \cong \mathbb{Z}$ generated by the vertical S^1-action.
(b) C is not central and $\pi_1(\mathrm{diff}_v(\Sigma))$ is trivial.

Suppose first that the fiber C is central in $\pi_1(\Sigma)$. Then there is a vertical S^1-action on Σ which moves the basepoint (in C) once around C. This maps onto the generator of $\pi_1(\mathrm{diff}(C))$, so $\pi_1(\mathrm{diff}_v(\Sigma)) \to \pi_1(\mathrm{diff}(C))$ is an isomorphism. Therefore $\pi_1(\mathrm{diff}_v(\Sigma))$ is infinite cyclic, with generator represented by the vertical S^1-action.

If the fiber is not central, then $\pi_1(\mathrm{diff}(C)) \to \pi_0(\mathrm{Diff}(\Sigma \text{ rel } C) \cap \mathrm{diff}_v(\Sigma))$ carries the generator to a diffeomorphism of Σ which induces an inner automorphism of infinite order on $\pi_1(\Sigma, x_0)$, where x_0 is a basepoint in C. Since elements of $\mathrm{Diff}(\Sigma \text{ rel } C)$ fix the basepoint, this diffeomorphism (and its powers) are not in $\mathrm{diff}(\Sigma \text{ rel } C)$. Therefore $\pi_1(\mathrm{diff}(C)) \to \pi_0(\mathrm{Diff}(\Sigma \text{ rel } C) \cap \mathrm{diff}_v(\Sigma))$ is injective, so $\pi_1(\mathrm{diff}_v(\Sigma))$ is trivial.

Now let $\mathcal{O}$ be the quotient orbifold, and consider the fibration of Theorem 3.9:

$$\mathrm{Diff}_v(\Sigma) \cap \mathrm{diff}_f(\Sigma) \longrightarrow \mathrm{diff}_f(\Sigma) \longrightarrow \mathrm{diff}(\mathcal{O})\,. \qquad (*)$$

Observe that $\mathrm{diff}(\mathcal{O})$ is homotopy equivalent to the identity component of the space of diffeomorphisms of the two-manifold $\mathcal{O} - \mathcal{E}$, where $\mathcal{E}$ is the exceptional set. Since Σ is Haken, this two-manifold is either a torus, annulus, disc with one puncture, Möbius band, or Klein bottle, or a surface of negative Euler characteristic. Therefore $\mathrm{diff}(\mathcal{O})$ is contractible unless $\chi(\mathcal{O} - \mathcal{E}) = 0$, in which case $\mathcal{E}$ is empty and $\mathcal{O}$ is an annulus or torus. Thus the higher homotopy groups of $\mathrm{diff}(\mathcal{O})$ are all trivial, and its fundamental group is isomorphic to the center of $\pi_1(\mathcal{O})$. When this center is nontrivial, the elements of $\pi_1(\mathcal{O})$ are classified by their traces at a basepoint of $\mathcal{O}$. From the exact sequence for the fibration $(*)$, it follows that $\pi_q(\mathrm{diff}_f(\Sigma)) = 0$ for $q \geq 2$.

To complete the proof, we recall the result of Hatcher [22] and Ivanov [32]: for M Haken, $\pi_q(\mathrm{diff}(M))$ is 0 for $q \geq 2$ and is isomorphic to the center of $\pi_1(M)$ for $q = 1$, and the elements of $\pi_1(\mathrm{diff}(M))$ are classified by their traces at the basepoint. We already have $\pi_q(\mathrm{diff}_f(\Sigma)) = 0$ for $q \geq 2$, so it remains to show that $\pi_1(\mathrm{diff}_f(\Sigma)) \to \pi_1(\mathrm{diff}(\Sigma))$ is an isomorphism.

Case I. $\pi_1(\mathcal{O})$ is centerless.

In this case diff($\mathcal{O}$) is contractible, and either C generates the center or $\pi_1(\Sigma)$ is centerless. From the exact sequence associated to the fibration (*), $\pi_1(\mathrm{diff}_v(\Sigma)) = \pi_1(\mathrm{Diff}_v(\Sigma) \cap \mathrm{diff}_f(\Sigma)) \to \pi_1(\mathrm{diff}_f(\Sigma))$ is an isomorphism. Suppose that C generates the center. Since $\pi_1(\mathrm{diff}_v(\Sigma))$ is infinite cyclic generated by the vertical S^1-action, Hatcher's theorem shows that the composition

$$\pi_1(\mathrm{diff}_v(\Sigma)) \to \pi_1(\mathrm{diff}_f(\Sigma)) \to \pi_1(\mathrm{diff}(\Sigma))$$

is an isomorphism. Therefore $\pi_1(\mathrm{diff}_f(\Sigma)) \to \pi_1(\mathrm{diff}(\Sigma))$ is an isomorphism. If $\pi_1(\Sigma)$ is centerless, then $\pi_1(\mathrm{diff}(\Sigma)) = 0$, $\pi_1(\mathrm{diff}_f(\Sigma)) \cong \pi_1(\mathrm{diff}_v(\Sigma)) = 0$, and again $\pi_1(\mathrm{diff}_f(\Sigma)) \to \pi_1(\mathrm{diff}(\Sigma))$ is an isomorphism.

Case II. $\pi_1(\mathcal{O})$ has nontrivial center.

Assume first that $\mathcal{O}$ is a torus. If Σ is the three-torus, then by considering the exact sequence for the fibration (*), one can check directly that the homomorphism $\partial\colon \pi_1(\mathrm{diff}(\mathcal{O})) \to \pi_0(\mathrm{Diff}_v(\Sigma) \cap \mathrm{diff}_f(\Sigma))$ is the zero map. We obtain the exact sequence

$$0 \longrightarrow \mathbb{Z} \longrightarrow \pi_1(\mathrm{diff}_f(\Sigma)) \longrightarrow \mathbb{Z}\times\mathbb{Z} \longrightarrow 0\,.$$

Since $\mathrm{diff}_f(\Sigma)$ is a topological group, $\pi_1(\mathrm{diff}_f(\Sigma))$ is abelian and hence isomorphic to $\mathbb{Z}\times\mathbb{Z}\times\mathbb{Z}$. The traces of the generating elements generate the center of $\pi_1(\Sigma)$, which shows that $\pi_1(\mathrm{diff}_f(\Sigma)) \to \pi_1(\mathrm{diff}(\Sigma))$ is an isomorphism.

Suppose that Σ is not a three-torus. Then Σ is a nontrivial S^1-bundle over $\mathcal{O}$, $\pi_1(\Sigma) = \langle a, b, t \mid tat^{-1} = a, [a,b] = 1, tbt^{-1} = a^n b\rangle$ for some integer n, and the fiber a generates the center of $\pi_1(\Sigma)$.

Let b_0 and t_0 be the image of the generators of b and t respectively in $\pi_1(\mathcal{O})$. Now $\pi_1(\mathrm{diff}(\mathcal{O})) \cong \mathbb{Z}\times\mathbb{Z}$ generated by elements whose traces represent the elements b_0 and t_0. By lifting these isotopies we see that $\partial\colon \pi_1(\mathrm{diff}(\mathcal{O})) \to \pi_0(\mathrm{diff}_v(\Sigma))$ is injective. Therefore $\pi_1(\mathrm{diff}_v(\Sigma))$ is isomorphic to $\pi_1(\mathrm{diff}_f(\Sigma))$, and the result follows as in case I.

Assume now that $\mathcal{O}$ is a Klein bottle. The Σ is an S^1-bundle over $\mathcal{O}$, $\pi_1(\Sigma) = \langle a, b, t \mid tat^{-1} = a^{-1}, [a,b] = 1, tbt^{-1} = a^{-n}b^{-1}\rangle$ for some integer n, with fiber a, and $\pi_1(\mathcal{O}) = \langle b_0, t_0 \mid t_0 b_0 t_0^{-1} = b_0^{-1}\rangle$. Now $\pi_1(\mathrm{diff}(\mathcal{O}))$ is generated by an isotopy whose trace represents the generator of the center of $\pi_1(\mathrm{diff}(\mathcal{O}))$, the element t_0^2. Observe that $\pi_1(\Sigma)$ has center if and only if $n = 0$.

If $n = 0$, then it follows that $\partial\colon \pi_1(\mathrm{diff}(\mathcal{O})) \to \pi_0(\mathrm{Diff}_v(\Sigma) \cap \mathrm{diff}_f(\Sigma))$ is the zero map. Hence $\pi_1(\mathrm{diff}_f(\Sigma)) \to \pi_1(\mathrm{diff}(\mathcal{O}))$ is an isomorphism and the generator of $\pi_1(\mathrm{diff}_f(\Sigma))$ is represented by an isotopy whose trace represents the element t^2. By Hatcher's result, $\pi_1(\mathrm{diff}_f(\Sigma)) \to \pi_1(\mathrm{diff}(\Sigma))$ is an isomorphism.

If $n \neq 0$, then $\partial\colon \pi_1(\mathrm{diff}(\mathcal{O})) \to \pi_0(\mathrm{Diff}_v(\Sigma) \cap \mathrm{diff}_f(\Sigma))$ is injective. Since $\pi_1(\Sigma)$ is centerless, $\pi_1(\mathrm{Diff}_v(\Sigma) \cap \mathrm{diff}_f(\Sigma)) = 0$. This implies that $\pi_1(\mathrm{diff}_f(\Sigma)) = 0$, and again Hatcher's result applies.

The cases where $\mathcal{O}$ is an annulus, disc with one puncture, or a Möbius band are similar to those of the torus and Klein bottle. □

3.10 The Parameterized Extension Principle

As a final application of the methods of this section, we present a result which will be used, often without explicit mention, in our later work. For a parameterized family of diffeomorphisms $F: M \times W \to M$, we denote the restriction $F: M \times \{u\} \to M$ by $F_u \in \text{Diff}(M)$. By a *deformation* of a parameterized family of diffeomorphisms $F: M \times W \to M$, we mean a homotopy from F to a parameterized family $G: M \times W \to M$ of diffeomorphisms when F and G are regarded as maps from W to $\text{Diff}(M, M)$.

Theorem 3.16 (Parameterized Extension Principle). *Let M and W be compact smooth manifolds, let M_0 be a submanifold of M, and let U be an open subset of M with $M_0 \subset U$. Suppose that $F: M \times W \to M$ is a parameterized family of diffeomorphisms of M. If $g \in C^\infty((M_0, M_0 \cap \partial M) \times W, (M, \partial M))$ is sufficiently close to $F|_{M_0 \times W}$, then there is a deformation G of F such that $G|_{M_0 \times W} = g$, and $G = F$ on $(M - U) \times W$. By selecting g sufficiently close to $F|_{M_0 \times W}$, G may be selected arbitrarily close to F.*

Proof. We may assume that each F_u is the identity on M. Provided that g is sufficiently close to $F|_{M_0 \times W}$, the Logarithm Lemma 3.1 gives sections $X(g_u) \in \mathscr{X}(M_0, TM)$ such that $\text{Exp}(X(g_u))(x) = g_u(x)$. Applying the Extension Lemma 3.2 gives a continuous linear map $k: \mathscr{X}(M_0, TM) \to \mathscr{X}(M, TM)$ with $k(X)(x) = X(x)$ for $x \in M_0$. Finally, the Exponentiation Lemma 3.3 shows that for g in some neighborhood U of the inclusion family (that is, the parameterized family with each g_u the inclusion of M_0 into M), each $\text{TExp} \circ k \circ X$ carries U into parameterized families of diffeomorphisms. By local convexity, after making U smaller, if necessary, the resulting diffeomorphisms G_u will be isotopic to the original F_u by moving along the unique geodesic between $G_u(x)$ and $F_u(x)$, giving the required deformation. □

Chapter 4
Elliptic Three-Manifolds Containing One-Sided Klein Bottles

In this chapter, we will prove Theorem 1.2. Section 4.1 gives a construction of the elliptic three-manifolds that contain a one-sided geometrically incompressible Klein bottle; they are described as a family of manifolds $M(m,n)$ that depend on two integer parameters $m, n \geq 1$. Section 4.2 is a section-by-section outline of the entire proof, which constitutes the remaining sections of the chapter.

4.1 The Manifolds $M(m,n)$

Let K_0 be a Klein bottle, which will later be the special "base" Klein bottle in $M(m,n)$, and write $\pi_1(K_0) = \langle\, a,\, b \mid bab^{-1} = a^{-1} \,\rangle$. The four isotopy (as well as homotopy) classes of unoriented essential simple closed curves on K_0 are b, ab, a, and b^2, with b and ab orientation-reversing and a and b^2 orientation-preserving.

Let P be the I-bundle over K_0 with orientable total space. The free abelian group $\pi_1(\partial P)$ is generated by elements homotopic in P to a and b^2.

Let R be a solid torus containing a meridional 2-disk with boundary C, a circle in ∂R. For a pair (m,n) of relatively prime integers, the three-manifold $M(m,n)$ is formed by identifying ∂R and ∂P in such a way that C is attached along a simple closed curve representing the element $a^m b^{2n}$. If $m = 0$, the resulting manifold is $\mathbb{RP}^3 \# \mathbb{RP}^3$, while if $n = 0$ it is $S^2 \times S^1$. In these cases K_0 is compressible, so from now on we will assume that neither m nor n is zero. Since $M(-m,n) = M(m,n)$ and $M(-m,-n) = M(m,n)$, we can and always will assume that both m and n are positive.

Each fibering of K_0 extends to a Seifert fibering of $M(m,n)$, for which P and R are fibered submanifolds. If K_0 has the longitudinal fibering, then in ∂P the fiber represents b^2. The meridian circle C of R equals $ma+nb^2$. Choosing p and q so that $mp - nq = 1$, the element $L = qa + pb^2$ is a longitude of R, since the intersection number $C \cdot L = mp - nq = 1$. We find that $b^2 = mL - qC$, so on R the Seifert fibering has an exceptional fiber of order m, unless $m = 1$. If instead K_0 has the

S. Hong et al., *Diffeomorphisms of Elliptic 3-Manifolds*, Lecture Notes in Mathematics 2055, DOI 10.1007/978-3-642-31564-0_4,

meridional fibering, then the fiber represents a in ∂R, and since $a = pC - nL$, R has an exceptional fiber of order n, unless $n = 1$. In terms of m and n, then, the cases discussed in Sect. 1.2 are as follows: I is $m > 1$ and $n > 1$, II is $m = 1$ and $n > 1$, III is $m > 1$ and $n = 1$, and IV is $m = n = 1$.

The fundamental group of $M(m,n)$ has a presentation

$$\langle\, a,\, b \mid bab^{-1} = a^{-1},\, a^m b^{2n} = 1 \,\rangle\,.$$

Note that $a^{2m} = 1$ and $b^{4n} = 1$.

If n is odd, then $\pi_1(M(m,n)) \cong C_n \times D^*_{4m}$, where C_n is cyclic and

$$D^*_{4m} = \langle x, y | x^2 = y^m = (xy)^2 \rangle$$

is the binary dihedral group. The C_n factor is generated by b^4 and the D^*_{4m} factor by $x = b^n$ and $y = a$.

If n is even, write $C_{4n} = \langle t | t^{4n} = 1 \rangle$. Let Δ be the diagonal subgroup of index 2 in $C_{4n} \times D^*_{4m}$. That is, there is a unique homomorphism from C_{4n} onto C_2, and, since m is odd, a unique homomorphism from D^*_{4m} onto C_2. The latter sends y to 1. Combining these homomorphisms sends $C_{4n} \times D^*_{4m}$ onto C_2 with kernel Δ. The element (t^{2n}, y^m) is a central involution in Δ, and $\pi_1(M(m,n))$ is isomorphic to $\Delta / \langle (t^{2n}, y^m) \rangle$. The correspondence is that $a = (1, y)$ and $b = (t, x)$.

When $m = 1$, the groups reduce in both cases to a cyclic group of order $4n$. From [7] or [56], $M(1,n) = L(4n, 2n-1)$. This homeomorphism can be seen directly as follows. Let T be a solid torus with $H_1(\partial T)$ the free abelian group generated by λ, a longitude, and μ, the boundary of a meridian disk. Let C_1 and C_2 be disjoint loops in ∂T, each representing $2\lambda + \mu$. There is a Möbius band M in T with boundary C_2. The double of T is an $S^2 \times S^1$ in which M and the other copy of M form a one-sided Klein bottle. The double has a Seifert fibering which is longitudinal on the Klein bottle, nonsingular on its complement, and in which C_1 is a fiber. If the attaching map in the doubling is changed by Dehn twists about C_1, the resulting manifolds are of the form $M(1,n)$, since they still have fiberings which are longitudinal on the Klein bottle and nonsingular on its complement. Since μ intersects C_1 twice, the image of μ under k Dehn twists about C_1 is $\mu + 2k(\mu + 2\lambda) = 4k\lambda + (2k+1)\mu$, so the resulting manifold is $L(4k, 2k+1) = L(4k, 2k-1)$. It must equal $M(1,k)$ since $M(1,k)$ is the only manifold of the form $M(1,n)$ with fundamental group C_{4k}.

As we have seen, with the longitudinal fibering the manifolds $M(m,n)$ have fibers of orders 2, 2, and m, so in the terminology of [46], $M(2,n)$ is a quaternionic manifold, while for $m > 2$, $M(m,n)$ is a (nonquaternionic) prism manifold.

4.2 Outline of the Proof

By Theorem 1.1, the inclusion $\mathrm{Isom}(M(m,n)) \to \mathrm{Diff}(M(m,n))$ is a bijection on path components, so we need only prove that the inclusion $\mathrm{isom}(M(m,n)) \to \mathrm{diff}(M(m,n))$ of the connected components of the identity induces isomorphisms on all homotopy groups. The rest of this chapter establishes this when at least one of m or n is greater than 1, that is, for Cases I, II, and III in Sect. 1.2. The remaining possibility $M(1,1)$ is the lens space $L(4,1)$, for which the Smale Conjecture holds by Theorem 1.3 proven in Chap. 5.

In Sect. 4.3, we give a calculation of the connected components of the identity in the isometry groups of the $M(m,n)$, in the process establishing the viewpoint and notation needed in Sect. 4.4.

The first task in Sect. 4.4 is to observe that the elements of $\pi_1(M(m,n))$ preserve the fibers of the Hopf fibering of S^3. Consequently there is an induced Seifert fibering of the $M(m,n)$, which we call the Hopf Seifert fibering of $M(m,n)$. A certain torus T_0 in S^3, vertical in the Hopf fibering, descends to a vertical Klein bottle K_0 in $M(m,n)$ which we call the base Klein bottle. On K_0, the Hopf fibering of $M(m,n)$ restricts to the longitudinal fibering in Cases I and II and the meridional fibering in Case III. In Sect. 4.4, we also check that the isometries of $M(m,n)$ are fiber-preserving and act isometrically on the quotient orbifold.

Most of Sect. 4.4 is devoted to verifying two facts:

(a) The map from $\operatorname{isom}(M(m,n))$ to the space of fiber-preserving isometric embeddings of K_0 into $M(m,n)$, defined by restriction to K_0, is a homeomorphism onto the connected component of the inclusion (Lemma 4.1).
(b) The inclusion of the latter space into the space of all fiber-preserving embeddings of K_0 into $M(m,n)$ that are isotopic to the inclusion is a homotopy equivalence (Lemma 4.2).

The big picture of what is going on here can be seen by consideration of the three types of quotient orbifolds shown in Table 4.2, which correspond to Cases I, II, and III respectively. For the first two types, K_0 is the inverse image of a geodesic arc connecting the two order 2 cone points, and for the third type, K_0 is the inverse image of a "great circle" geodesic in $\mathbb{RP}^2$. The inverse images of such geodesics are the images of the fiber-preserving isometric embeddings isotopic to the inclusion, the so-called "special" Klein bottles. They are the translates of K_0 under $\operatorname{isom}(M(m,n))$ (which also contains "vertical" isometries that take each fiber to itself, so preserve each special Klein bottle). Our precise description of $\operatorname{isom}(M(m,n))$ allows us to examine its effects on these Klein bottles and establish fact (a). For the first two types of quotient orbifold, a fiber-preserving embedding of K_0 that is isotopic to the inclusion carries K_0 onto the inverse image of an arc connecting two order 2-cone points and isotopic (avoiding the third cone point, if there is one) to a geodesic arc, and for the third type they carry K_0 onto the inverse image of an essential circle in $\mathbb{RP}^2$. Fact (b) for the third type of orbifold boils down to the fact that the space of all essential embeddings of the circle in $\mathbb{RP}^2$ is homotopy equivalent to the space of geodesic embeddings (which is $L(4,1)$), and analogous properties of arcs in the other two types of orbifolds.

The reader who is comfortable with this summary of Sects. 4.3 and 4.4 has little need to wade through their details.

The Smale Conjecture for the $M(m,n)$ reduces to Theorem 4.1, which says that the inclusion of the space of *fiber-preserving* embeddings of K_0 into $M(m,n)$ into the space of *all* embeddings of K_0 into $M(m,n)$ is a homotopy equivalence (on the connected components of the inclusion $K_0 \to M(m,n)$). This reduction is the main content of Sect. 4.5, and is carried out using the results of Sect. 4.4 and routine manipulation of exact sequences arising from fibrations of various spaces of mappings.

The final three sections are the proof of Theorem 4.1. One must start with a family of embeddings of K_0 into $M(m,n)$ parameterized by D^k, and change it by homotopy as an element of $\mathrm{Maps}(D^k, \mathrm{emb}(K_0, M(m,n)))$ to a family of fiber-preserving embeddings. The embeddings are fiber-preserving at parameters in ∂D^k, and this property must be retained so during the homotopy. Sections 4.6 and 4.7 are auxiliary results needed for the main argument in Sect. 4.8.

In Sect. 4.6, we analyze the situation when an embedded Klein bottle K meets K_0 in "generic position," meaning that all tangencies are of finite multiplicity type. In $M(m,n)$, K_0 has a standard neighborhood which is a twisted I-bundle P, and $P - K_0$ has a product structure $T \times (0,1]$ with each $T_u = T \times \{u\}$ a fibered "level" torus. The key result of the analysis is Proposition 4.1, which says for all u sufficiently close to 0, each circle of $K \cap T_u$ is either inessential in T_u, or represents a or b^2 in $\pi_1(T_u)$. We will see below where this critical fact is needed.

The proof of Proposition 4.1 uses a technique which may seem surprising in our differentiable context. Since we may not have full transversality, we go ahead and make the situation much less transverse, by a process called *flattening*. It moves K to a PL-embedded Klein bottle that intersects K_0 in a 2-complex, but still meets torus levels for u near 0 in loops isotopic to their original intersection circles with K. For these flattened surfaces, combinatorial arguments can be used to establish that those intersection circles are a- and b^2-curves. Proposition 4.1 fails for $M(1,1)$, as we show by example.

Section 4.7 recalls Ivanov's idea [36] of perturbing a parameterized family of embeddings of K_0 into $M(m,n)$ so that each image meets K_0 in generic position. A bit of extra work is needed to ensure that during a homotopy from our original family to the generic position family, the embeddings remain fiber-preserving at parameters in ∂D^k.

Section 4.8 is the argument to make a parameterized family of embeddings $K_0 \to K_t \subset M(m,n), t \in D^k$, fiber-preserving for the Hopf fibering on $M(m,n)$. The first step is a minor technical trick needed to ensure that no K_t equals K_0; this allows Sect. 4.7 to be applied to assume that the K_t meet K_0 in generic position. Next, we use Hatcher's methods to simplify the intersections of the Klein bottles K_t with the torus levels T_u. Each K_t has finitely many associated torus levels T_u, obtained using Proposition 4.1. First, we eliminate intersections that are contractible in K_t (and hence in T_u). This part of the argument, called Step 2, is a straightforward adaptation of Hatcher's arguments from [23, 25], but we give a fair amount of detail since these methods are not widely used.

Step 3 is where the hard work from Sect. 4.6 comes into play. From our analysis of generic position configurations, specifically Proposition 4.1, we know that K_t

meets its associated levels T_u in circles that represent a or b^2 in $\pi_1(T_u)$. Now, T_u separates $M(m,n)$ into a twisted I-bundle P_u, containing K_0, and a solid torus R_u. Some homological arguments (which again break down for $M(1,1)$) show that a circle of $K_t \cap T_u$ is a longitude of R_u only if it is isotopic in T_u to a fiber. Hence any circles of $K_t \cap T_u$ that are not isotopic to fibers are also not longitudes of R_u, and consequently the annuli of $K_t \cap R_u$ that contain them are uniquely boundary-parallel in R_u. This allows us to once again apply Hatcher's parameterized methods to pull the annuli of $K_t \cap R_u$ whose boundary circles are not isotopic in T_u to fibers out of R_u, achieving that every loop of $K_t \cap T_u$ is isotopic in T_u to a fiber.

Two tasks remain:

1. Make K_t intersect its associated levels T_u in circles that are fibers and are the images of fibers of K_0 under the embedding $K_0 \to M(m,n)$.
2. Make the embeddings fiber-preserving on the intersections of K_t with the other pieces of $M(m,n)$, which are topologically either twisted I-bundles over K_0, product regions between levels, or solid tori that are complements of twisted I-bundles over K_0.

The underlying facts about fiber-preserving embeddings needed for this are given in Step 4. The final part of the argument, Step 5, applies these facts, working up the skeleta of a triangulation of D^k, to complete the deformation.

4.3 Isometries of Elliptic Three-Manifolds

In Sect. 1.1, we recalled the isometry groups of elliptic three-manifolds. We will now present the calculations of these groups—actually, only the connected component isom(M) of the identity—for the elliptic three-manifolds that contain a geometrically incompressible Klein bottle. Besides giving an opportunity to revisit the beautiful interaction between the structure of S^3 as the unit quaternions and the structure of SO(4), which will provide the setting for some key technical results in Sect. 4.4.

Fix coordinates on S^3 as $\{(z_0, z_1) \mid z_i \in \mathbb{C}, z_0\overline{z_0} + z_1\overline{z_1} = 1\}$. Its group structure as the unit quaternions can then be given by writing points in the form $z_0 + z_1 j$, where $j^2 = -1$ and $j z_i = \overline{z_i}\, j$. The unique element of order 2 in S^3 is -1, and it generates the center of S^3.

By S^1 we will denote the subgroup of points in S^3 with $z_1 = 0$, that is, all quaternions of the form z_0, where z_0 lies in the unit circle in $\mathbb{C}$. Let $\xi_k = \exp(2\pi i/k)$, which generates a cyclic subgroup $C_k \subset S^1$. The elements $S^1 \cup S^1 j$ form a subgroup $\mathrm{O}(2)^* \subset S^3$, which is exactly the normalizer of C_k if $k > 2$. Also contained in $\mathrm{O}(2)^*$ is the binary dihedral group D^*_{4m} generated by $x = j$ and $y = \xi_{2m}$; its normalizer is D^*_{8m}. By J we denote the subgroup of S^3 consisting of the elements with both z_0 and z_1 real. It is the centralizer of j.

The real part $\Re(z_0+z_1 j)$ is the real part $\Re(z_0)$ of the complex number z_0, and the imaginary part $\Im(z_0+z_1 j)$ is $\Im(z_0)+z_1 j$. The usual inner product on S^3 is given by $z \cdot w = \Re(zw^{-1})$, where $\Re(z_0+z_1 j) = \Re(z_0)$. Consequently, left multiplication and right multiplication by elements of S^3 are orthogonal transformations of S^3, and there is a homomorphism $F\colon S^3 \times S^3 \to \mathrm{SO}(4)$ defined by $F(q_1, q_2)(q) = q_1 q q_2^{-1}$. It is surjective and has kernel $\{(1,1), (-1,-1)\}$.

The quaternions with real part 0 are the pure imaginary quaternions, and form a subspace $P \subset S^3$ homeomorphic to S^2. In fact, P is exactly the orthogonal complement of 1. Conjugation by elements of S^3 preserves P, defining a surjective homomorphism $S^3 \to \mathrm{SO}(3)$ with kernel $\langle \pm 1 \rangle$.

Suppose that G is a finite subgroup of SO(4) acting freely on S^3. Since SO(4) is the full group of orientation-preserving isometries of S^3, the orientation-preserving isometries $\mathrm{Isom}_+(S^3/G)$ are the quotient $\mathrm{Norm}(G)/G$, where $\mathrm{Norm}(G)$ is the normalizer of G in SO(4). Assuming that the group G is clear from the context, we denote the isometry that an element $F(q_1, q_2)$ of $\mathrm{Norm}(G)$ induces on S^3/G by $f(q_1, q_2)$.

Let $G^* = F^{-1}(G)$, and let G_L and G_R be the projections of G^* into the left and right factors of $S^3 \times S^3$. Notice that $\mathrm{Norm}(G)/G \cong \mathrm{Norm}(G^*)/G^*$. The connected component of the identity in $\mathrm{Norm}(G^*)$ is denoted by $\mathrm{norm}(G^*)$. Since G^* is discrete, these elements centralize G^*. Consequently, $\mathrm{norm}(G^*)$ is the product $\mathrm{norm}(G_L) \times \mathrm{norm}(G_R)$ of the corresponding connected normalizers of G_L and G_R in the S^3 factors. The connected component of the identity in the isometry group of S^3/G is then $\mathrm{isom}(M) = \mathrm{norm}(G^*)/(G^* \cap \mathrm{norm}(G^*))$. We now compute $\mathrm{isom}(M(m,n))$ for the four cases listed in Sect. 1.2:

Cases II and IV. $m = 1$.

The element $F(\xi_{4n}^{n-1}, i)$ acts on S^3 by

$$F(\xi_{4n}^{n-1}, i)(z_0 + z_1 j) = \xi_{4n}^{n-1} z_0(-i) + \xi_{4n}^{n-1} z_1 j(-i) = \xi_{4n}^{-1} z_0 + \xi_{4n}^{2n-1} z_1 j\,.$$

Consequently the quotient of S^3 by the subgroup generated by $F(\xi_{4n}^{n-1}, i)$ is $L(4n, 2n+1) = L(4n, 2n-1) = M(1,n)$. For some work in Sect. 4.4, however, it is more convenient to use a conjugate of this generator. Conjugation by $F(1, \frac{1}{\sqrt{2}} i + \frac{1}{\sqrt{2}} j)$ moves $F(\xi_{4n}^{n-1}, i)$ to $F(\xi_{4n}^{n-1}, j)$. The latter will be our standard generator for $G = \pi_1(M(1,n))$.

Letting G be the group C_{4n} generated by $F(\xi_{4n}^{n-1}, j)$, G_R is the cyclic subgroup of order 4 generated by j, so $\mathrm{norm}(G_R) = J$, and G_L is generated by $\{\xi_{4n}^{n-1}, -1\}$. If $n = 1$, then $\xi_{4n}^{n-1} = 1$ and $G_L = C_2$. If $n > 1$ then ξ_{4n}^{n-1} has order $4n/\gcd(4n, n-1) = 4n/\gcd(4, n-1)$, so G_L is C_{4n} if n is even, C_{2n} if $n \equiv 3 \mod 4$, and C_n if $n \equiv 1 \mod 4$.

1. If $n = 1$, then $\mathrm{norm}(G_L) = S^3$, and $\mathrm{isom}(M(1,1)) \cong \mathrm{SO}(3) \times S^1$, consisting of all isometries of the form $f(q,x)$ with $(q,x) \in S^3 \times J$.
2. If $n > 1$, then $\mathrm{norm}(G_L) = S^1$, so $\mathrm{isom}(M(1,n)) \cong S^1 \times S^1$, consisting of all isometries of the form $f(x_1, x_2)$ with $(x_1, x_2) \in S^1 \times J$.

Table 4.1 Isometry groups of the $M(m,n)$

m, n values	M	$\mathrm{Isom}(M)$
$m = n = 1$	$L(4,1)$	$\mathrm{SO}(3)\times S^1 = \{f(q,x) \mid (q,x) \in S^3 \times J\}$
$m = 1, n > 1$	$L(4n, 2n-1)$	$S^1 \times S^1 = \{f(x,y) \mid (x,y) \in S^1 \times J\}$
$m > 1, n = 1$	Quaternionic ($m = 2$) or prism ($m > 2$)	$\mathrm{SO}(3) = \{f(1,q) \mid q \in S^3\}$
$m > 1, n > 1$	Quaternionic ($m = 2$) or prism ($m > 2$)	$S^1 = \{f(x,1) \mid x \in S^1\}$

Case III. $m > 1$ and $n = 1$.

We embed $G = D^*_{4m}$ in SO(4) as the subgroup $F(D^*_{4m} \times \{1\})$. We have $G_L = D^*_{4m}$ and $G_R = C_2$, so $\mathrm{norm}(G_L) \times \mathrm{norm}(G_R) = \{1\} \times S^3$. Therefore $\mathrm{isom}(M(m,1)) \cong \mathrm{SO}(3)$, and consists of all isometries of the form $f(1,q)$.

Case I. $m > 1$ and $n > 1$.

If n is odd, then $G = C_n \times D^*_{4m}$, and we embed G in SO(4) as $F(C_{2n} \times D^*_{4m})$, so $G_L = C_{2n}$ and $G_R = D^*_{4m}$. If n is even, then G is the image in SO(4) of the unique diagonal subgroup of index 2 in $C_{4n} \times D^*_{4m}$, so $G_L = C_{4n}$ and $G_R = D^*_{4m}$. In either case, we have $\mathrm{norm}(G_L) \times \mathrm{norm}(G_R) = S^1 \times \{1\}$. Therefore $\mathrm{isom}(M(m,n)) \cong S^1$, and consists of all isometries of the form $f(x,1)$ with $x \in S^1$.

Table 4.1 summarizes our calculations of $\mathrm{Isom}(M(m,n))$.

4.4 The Hopf Fibering of $M(m,n)$ and Special Klein Bottles

From now on, we use M to denote one of the manifolds $M(m,n)$ with $m > 1$ or $n > 1$. In this section, we construct certain Seifert fiberings of these M, which we will call their Hopf fiberings, and examine the effect of $\mathrm{isom}(M)$ on them. Also, we define certain vertical Klein bottles in M, called special Klein bottles, are deeply involved in the reductions carried out in Sect. 4.5. A certain special Klein bottle K_0, called the base Klein bottle, will play a key role.

We will regard the 2-sphere S^2 as $\mathbb{C} \cup \{\infty\}$. We speak of antipodal points and orthogonal transformations on S^2 by transferring them from the unit 2-sphere using the stereographic projection that identifies the point (x_1, x_2, x_3) with $(x_1 + x_2 i)/(1 - x_3)$. For example, the antipodal map α is defined by $\alpha(z) = -1/\overline{z}$.

As is well-known, the Hopf fibering on S^3 is an S^1-bundle structure with projection map $H\colon S^3 \to S^2$ defined by $H(z_0, z_1) = z_0/z_1$. The left action of S^1 on S^3 takes each Hopf fiber to itself, so preserves the Hopf fibering. The element $F(j,1)$ also preserves it. For $j(z_0 + z_1 j) = -\overline{z_1} + \overline{z_0} j$, so $H(F(j,1)(z_0 + z_1 j)) = -1 / \overline{z_0/z_1}$. Right multiplication by elements of S^3 commutes with the left action of S^1, so it preserves the Hopf fibering, and there is an induced action of S^3 on S^2. In fact, it acts orthogonally. For if we write $x = x_0 + x_1 j$ and $z = z_0 + z_1 j$, we have $zx^{-1} = z_0\overline{x_0} + z_1\overline{x_1} + (z_1 x_0 - z_0 x_1) j$, so the induced action on S^2 carries

Table 4.2 Quotient orbifolds for the Hopf fiberings

m, n values	$h(\pi_1(M))$	$\mathcal{O}$	$\mathrm{Isom}(\mathcal{O})$
$m > 1, n > 1$	$D_{2m} = \langle r_m, t\rangle$	$(S^2; 2, 2, m)$	$\{1\}$
$m = 1, n > 1$	$C_2 = \langle t\rangle$	$(S^2; 2, 2)$	$\mathrm{SO}(2)$
$m > 1, n = 1$	$C_2 = \langle \alpha\rangle$	$(\mathbb{RP}^2;)$	$\mathrm{SO}(3)$

$$z_0/z_1 \text{ to } (z_0\overline{x_0} + z_1\overline{x_1})/(z_1x_0 - z_0x_1) = \frac{\overline{x_0}\left(\dfrac{z_0}{z_1}\right) + \overline{x_1}}{-x_1\left(\dfrac{z_0}{z_1}\right) + x_0} = \begin{pmatrix} \overline{x_0} & \overline{x_1} \\ -x_1 & x_0 \end{pmatrix}(z_0/z_1).$$

The trace of this linear fractional transformation is real and lies between -2 and 2 (unless $x = \pm 1$, which acts as the identity on S^2), so it is elliptic. Its fixed points are $\left((x_0 - \overline{x_0}) \pm \sqrt{(x_0 - \overline{x_0})^2 - 4x_1\overline{x_1}}\right)/(2x_1)$, which are antipodal, so it is an orthogonal transformation. Combining these observations, we see that the action induced on S^2 via H determines a surjective homomorphism $h\colon \mathrm{O}(2)^* \times S^3 \to \mathrm{O}(3)$, given by $h(x_0, 1) = 1$ for $x_0 \in S^1$, $h(j, 1) = \alpha$, and $h(1, x_0 + x_1 j) = \begin{pmatrix} \overline{x_0} & \overline{x_1} \\ -x_1 & x_0 \end{pmatrix}$. The kernel of h is $S^1 \times \{\pm 1\}$.

With the explicit embeddings selected in Sect. 4.3, each of our groups $G = \pi_1(M)$ lies in $F(\mathrm{O}(2)^* \times S^3)$, so preserves the Hopf fibering, and descends to a Seifert fibering on $M(m, n) = S^3/G$.

Definition 4.1. The *Hopf fibering* of $M(m, n)$ is the image of the Hopf fibering of S^3 under the quotient map $S^3 \to S^3/\pi_1(M(m, n))$. We will always use the Hopf fibering on the manifolds $M(m, n)$.

The Hopf fibering $H\colon S^3 \to S^2$ induces the orbit map $M(m, n) \to S^2/h(G)$, and the orbit map is induced by the composition of H followed by the quotient map from S^2 to the quotient orbifold $S^2/h(G)$. The quotient orbifolds for our fiberings are easily calculated using the explicit embeddings of G into $\mathrm{SO}(4)$ given in Sect. 4.3, together with the facts that $h(j, 1) = \alpha$, $h(1, \xi_{2m}) = r_m$, the (clockwise) rotation through an angle $2\pi/m$ with fixed points 0 and ∞, defined by $r_m(z) = \xi_m^{-1} z$, and $h(1, j) = t$, the rotation through an angle π with fixed points $\pm i$, defined by $t(z) = -1/z$. Table 4.2 lists the various cases, where $(F; n_1, \ldots, n_k)$ denotes the 2-orbifold with underlying topological space the surface F and k cone points of orders $n_1, \ldots, n_k$.

Since $m > 1$ or $n > 1$, we have $\mathrm{norm}(\pi_1(M)) \subset F(\mathrm{O}(2)^* \times S^3)$, so $\mathrm{isom}(M)$ preserves the Hopf fibering. Since the quotient orbifolds are the quotients of orthogonal actions on S^2, they have metrics of constant curvature 1, except at the cone points, where the cone angle at an order k cone point is $2\pi/k$. Table 4.2 shows the quotient orbifolds with shapes that suggest the symmetries for this constant curvature metric. The isometry group of each orbifold $\mathcal{O}$ is the normalizer of its orbifold fundamental group $h(G)$ in the isometry group $\mathrm{O}(3)$ of S^2. The homomorphism h induces a homomorphism $\mathrm{isom}(M) \to \mathrm{isom}(\mathcal{O})$, and from the explicit description of $\mathrm{isom}(M)$ from Table 4.1 we can use h to compute the image. In each case, all isometries in the connected component of the identity, $\mathrm{isom}(\mathcal{O})$, are induced by elements of $\mathrm{isom}(M)$. (The groups $\mathrm{isom}(\mathcal{O})$ are computed as $\mathrm{norm}(G)/(G \cap \mathrm{norm}(G))$ where $\mathrm{norm}(G)$ is the connected component of the identity in the normalizer of G in $\mathrm{isom}(S^2) = \mathrm{SO}(3)$. In particular, $\mathrm{isom}(\mathbb{RP}^2) = \mathrm{SO}(3)$, which can be seen directly by noting that each isometry of $\mathbb{RP}^2$ lifts to an unique orientation-preserving isometry of S^2.)

Our next task is to understand the fibered Klein bottles in M.

Definition 4.2. A torus $T \subset S^3$ *special* if its image in S^2 under H is a great circle. Klein bottles in M that are the images of special tori in S^3 are called *special Klein bottles*. A suborbifold in $\mathcal{O}$ is called *special* when it is either

(i) A one-sided geodesic circle (when $\mathcal{O} = \mathbb{RP}^2$), or
(ii) A geodesic arc connecting two order-2 cone points (in the other two cases).

Clearly special tori are vertical in the Hopf fibering. We remark that special tori are Clifford tori, that is, they have induced curvature zero in the usual metric on S^3.

A Klein bottle in M is special if and only if its image in $\mathcal{O}$ is a special suborbifold. To see this, consider a special torus T in S^3. If its image in $\mathcal{O}$ is special, then its image in M is a one-sided submanifold, so must be a Klein bottle. Conversely, the projection of T to $\mathcal{O}$ must always be a geodesic, and if its image in M is a submanifold, then the projection to $\mathcal{O}$ cannot have any self intersections or meet a cone point of order more than 2. And if the projection is a circle, it is one-sided if and only if the image of T in M is one-sided.

Note that the fibering on a special Klein bottle is meridional (i.e. an S^1-bundle over S^1) in case (i), and longitudinal (two exceptional fibers that are center circles of Möbius bands) in case (ii). From Table 4.2, we see that:

1. When $n = 1$, special Klein bottles have the meridional fibering.
2. When $n > 1$, special Klein bottles have the longitudinal fibering.

Let T_0 be the fibered torus $H^{-1}(U)$, where U is the unit circle in S^2. Explicitly, T_0 consists of all $z_0 + z_1 j$ for which $|z_0| = |z_1| = \frac{1}{\sqrt{2}}$. Observe that the isometries $F(\mathrm{O}(2)^* \times \mathrm{O}(2)^*)$ of S^3 leave T_0 invariant. The action of $F(\mathrm{O}(2)^* \times \mathrm{O}(2)^*)$ on T_0 can be calculated using the normalized coordinates $[x_0, y_0] \in S^1 \times S^1/\langle(-1,-1)\rangle$, where $[x_0, y_0]$ corresponds to the point $x_0\left(\frac{1}{\sqrt{2}}\,\overline{y_0}\,i + \frac{1}{\sqrt{2}}\,y_0\,j\right)$. For $(z_0, w_0) \in S^1 \times S^1$, we have $F(z_0, w_0)[x_0, y_0] = [z_0 x_0, w_0 y_0]$. Also:

(a) $F(j,1)[x_0, y_0] = [-\overline{x_0}, i\,y_0]$. Viewed in the fundamental domain $\Im(x_0) \geq 0$ for the involution on $T_0 = \{(x_0, y_0)\}$ that multiplies by $(-1,-1)$, this rotates the y_0-coordinate through $\pi/2$, and reflects in the x_0-coordinate fixing the point i.
(b) $F(1,j)[x_0, y_0] = [i x_0, -\overline{y_0}]$. Again viewing in the fundamental domain $\Im(y_0) \geq 0$ in T_0, this rotates the x_0-coordinate through $\pi/2$, and reflects in the y_0-coordinate fixing the point i.

In fact, the restriction of $F(\mathrm{O}(2)^* \times \mathrm{O}(2)^*)$ to T_0 is exactly the group of all fiber-preserving isometries $\mathrm{Isom}_f(T_0)$. The Hopf fibers are the orbits of the action of

$F(S^1 \times \{1\})$ on T_0, so are the circles with constant y_0-coordinate. Using (a) and (b), we find that $F(j,i)[x_0, y_0] = [\overline{x_0}, y_0]$ and $F(i,j)[x_0, y_0] = [x_0, \overline{y_0}]$. The elements $F(z_0, w_0)$ act transitively on T_0, and only the two reflections $F(i,j)$ and $F(j,i)$ and their composition fix $[1,1]$ and preserve the fibers, so together they generate all the fiber-preserving isometries.

Since $F(\mathrm{O}(2)^* \times \mathrm{O}(2)^*)$ contains (each of our groups) G, the image of T_0 in M is a fibered submanifold K_0. When $n = 1$, the image of T_0 in $\mathcal{O}$ is a geodesic circle which is the center circle of a Möbius band, and when $n > 1$ its image is a geodesic arc connecting two cone points of order 2, so K_0 is a special Klein bottle.

Definition 4.3. The special Klein bottle K_0 is called the *base Klein bottle* of $M(m,n)$.

Since K_0 is special, it has the meridional or longitudinal fibering according as $n = 1$ or $n > 1$.

Since G acts by isometries on S^3, the subspace metric on T_0 induces a metric on K_0 such that the inclusion of K_0 into M is isometric. Denote by $\mathrm{isom}_f(K_0, M)$ the connected component of the inclusion in the space of all fiber-preserving isometric embeddings of K_0 into M. Since the isometries of M are fiber-preserving, their compositions with the inclusion determine a map $\mathrm{isom}(M) \to \mathrm{isom}_f(K_0, M)$. By composition with the inclusion, we may regard $\mathrm{isom}_f(K_0)$, the connected component of the identity in the group of fiber-preserving isometries of K_0, as a subspace of $\mathrm{isom}_f(K_0, M)$.

Lemma 4.1. *If $m > 1$ or $n > 1$, then* $\mathrm{isom}(M) \to \mathrm{isom}_f(K_0, M)$ *is a homeomorphism. Moreover,*

(i) *If $n = 1$, then the elements $f(1, w_0)$ for $w_0 \in S^1$ preserve K_0, and restriction of this subgroup of* $\mathrm{isom}(M)$ *gives a homeomorphism* $S^1 \to \mathrm{isom}_f(K_0)$.
(ii) *If $n > 1$, then the elements $f(x_0, 1)$ for $x_0 \in S^1$ preserve K_0, and restriction of this subgroup of* $\mathrm{isom}(M)$ *gives a homeomorphism* $S^1 \to \mathrm{isom}_f(K_0)$.

Proof. For injectivity, suppose that an element of $\mathrm{isom}(M)$ fixes each point of K_0. Since it is isotopic to the identity, it cannot locally interchange the sides of K_0. Since it is an isometry, this implies it is the identity on all of M.

For surjectivity, we first examine the action of $\mathrm{isom}(M)$ on special Klein bottles in M.

For the quotient orbifolds of the form $(S^2; 2, 2)$, the special suborbifolds are the portions of great circles running between the two cone points, and for those of the form $(\mathbb{RP}^2;)$, they are the images of great circles under $S^2 \to \mathbb{RP}^2$. For those of the form $(S^2; 2, 2, m)$ with $m > 2$, the geodesic running between the two order-2 cone points is the unique special suborbifold. In all of these cases, $\mathrm{isom}(\mathcal{O})$ acts transitively on the special suborbifolds. In the remaining case of $(S^2; 2, 2, 2)$, there are three special suborbifolds corresponding to the three nonisotopic special Klein bottles in M, and $\mathrm{isom}(\mathcal{O})$ acts transitively on the special suborbifolds isotopic to K_0. Since all elements of $\mathrm{isom}(\mathcal{O})$ are induced by elements of $\mathrm{isom}(M)$, it follows that in all cases, $\mathrm{isom}(M)$ acts transitively on the space of special Klein bottles in M that are isotopic to K_0.

Since $\mathrm{isom}(M)$ acts transitively on the space of special Klein bottles isotopic to K_0, it remains to check that any element of $\mathrm{isom}_f(K_0, M)$ that carries K_0 to K_0 is the restriction of an element of $\mathrm{isom}(M)$.

Consider first the case when $m > 1$ and $n = 1$, so G is $F(D^*_{4m} \times \{1\})$ and K_0 has the meridional fibering. The fiber-preserving isometry group $\mathrm{Isom}_f(K_0)$ is $\mathrm{Norm}(G)/G$ where $\mathrm{Norm}(G)$ is the normalizer of G in $\mathrm{Isom}_f(T_0)$. The elements in $F(C_{2m} \times \{1\})$ rotate in the x_0-coordinate, while the element $F(j, 1)$ is as described in (a). So each element of $G - C_{2k}$ leaves invariant a pair of circles each having constant x_0-coordinate. The union of these invariant circles for all the elements of $G - C_{2k}$ must be invariant under the action of $\mathrm{Norm}(G)$ on T_0, so the identity component of $\mathrm{Norm}(G)$ consists only of $F(\{1\} \times S^1)$. Consequently the elements $f(1, w_0)$ of $\mathrm{isom}(M)$ induce all elements of $\mathrm{isom}_f(K_0)$, proving the surjectivity of $\mathrm{isom}(M) \to \mathrm{isom}_f(K_0, M)$ and verifying assertion (i).

For $m = 1$ and $n > 1$, G is cyclic generated by $F(\xi_{4n}^{n-1}, j)$ and K_0 has the longitudinal fibering. Since $F(1, j)$ is as described in (b), there is a pair of circles in T_0, each having constant y_0-coordinate and each invariant under all elements of G (these circles become the exceptional fibers in K_0). Since these circles must be invariant under the normalizer of G in $\mathrm{Isom}(T_0)$, the identity component of $\mathrm{Norm}(G)$ consists only of $F(S^1 \times \{1\})$. Therefore the isometries $f(x_0, 1)$ with $x_0 \in S^1$ of $\mathrm{isom}(M)$ induce all elements of $\mathrm{isom}_f(K_0)$, proving the surjectivity of $\mathrm{isom}(M) \to \mathrm{isom}_f(K_0, M)$ and verifying assertion (ii) for this case.

Finally, if both $m > 1$ and $n > 1$, then G contains $F(1, j)$ and K_0 has the longitudinal fibering. Again, the identity component of $\mathrm{Norm}(G)$ is $F(S^1 \times \{1\})$, and the isometries $f(x_0, 1)$ with $x_0 \in S^1$ induce all of $\mathrm{isom}_f(K_0)$. □

In the space of all smooth fiber-preserving embeddings of K_0 in M (for the appropriate fibering on K_0), let $\mathrm{emb}_f(K_0, M)$ denote the connected component of the inclusion.

Lemma 4.2. *If either $m > 1$ or $n > 1$, then the inclusion*

$$\mathrm{isom}_f(K_0, M) \to \mathrm{emb}_f(K_0, M)$$

is a homotopy equivalence.

Proof. Let $\mathscr{K}_0$ be the image of K_0 in the quotient orbifold $\mathscr{O}$ of the fibering on M. As we have seen, when K_0 has the meridional fibering, $\mathscr{K}_0$ is a one-sided geodesic circle in $\mathscr{O}$, and when K_0 has the longitudinal fibering, $\mathscr{K}_0$ is a geodesic arc connecting two order 2 cone points of $\mathscr{O}$. Let $\mathrm{emb}(\mathscr{K}_0, \mathscr{O})$ denote the connected component of the inclusion in the space of orbifold embeddings, and $\mathrm{isom}(\mathscr{K}_0, \mathscr{O})$ its subspace of isometric embeddings, and let a subscript v as in $\mathrm{Diff}_v(K_0)$ indicate the vertical maps—those that take each fiber to itself. Consider the following diagram, which we call the main diagram:

$$\begin{array}{ccccc}
\mathrm{Isom}_v(K_0) \cap \mathrm{isom}_f(K_0) & \longrightarrow & \mathrm{isom}_f(K_0, M) & \longrightarrow & \mathrm{isom}(\mathscr{K}_0, \mathscr{O}) \\
\downarrow & & \downarrow & & \downarrow \\
\mathrm{Diff}_v(K_0) \cap \mathrm{diff}_f(K_0) & \longrightarrow & \mathrm{emb}_f(K_0, M) & \longrightarrow & \mathrm{emb}(\mathscr{K}_0, \mathscr{O})
\end{array}$$

in which the vertical maps are inclusions. The left-hand horizontal arrows are inclusions, and the right-hand horizontal arrows take each embedding to the embedding induced on the quotient objects. By Theorem 3.11, the bottom row is a fibration. We will now examine the top row.

Suppose first that $n = 1$, so that $\mathscr{O} = (\mathbb{RP}^2;)$ and $\mathscr{K}_0$ is the image of the unit circle U of S^2. For this case, $\mathrm{isom}(\mathscr{K}_0, \mathscr{O})$ can be identified with the unit tangent space of $\mathbb{RP}^2$. For if we fix a unit tangent vector of $\mathscr{K}_0$, the image of this vector under an isometric embedding is a unit tangent vector to $\mathbb{RP}^2$, and each unit tangent vector of $\mathbb{RP}^2$ corresponds to a unique isometric embedding of $\mathscr{K}_0$. To understand this unit tangent space, note first that the unit tangent space of S^2 is $\mathbb{RP}^3$, since each unit tangent vector to S^2 corresponds to a unique element of $\mathrm{SO}(3) = \mathbb{RP}^3$. The unit tangent space of S^2 double covers the unit tangent space of $\mathbb{RP}^2$, so the latter must be $L(4, 1)$.

Since the isometries of M are all fiber-preserving, there is a commutative diagram

$$\begin{array}{ccc}
\mathrm{isom}(M) & \xrightarrow{\;\overline{h}\;} & \mathrm{isom}(\mathscr{O}) \\
{\scriptstyle \widetilde{\varrho}}\downarrow & & {\scriptstyle \varrho}\downarrow \\
\mathrm{isom}_f(K_0, M) & \longrightarrow & \mathrm{isom}(\mathscr{K}_0, \mathscr{O})
\end{array}$$

where $\overline{h}$ is induced by the homomorphism $h\colon \mathrm{O}(2)^* \times S^3 \to \mathrm{O}(3)$ defined near the beginning of this section. By Lemma 4.1, the restriction $\widetilde{\varrho}$ is a homeomorphism, and from Table 4.2, $\overline{h}$ is a homeomorphism. The restriction ϱ is a twofold covering map, since there are two isometries that restrict to the inclusion on $\mathscr{K}_0$: the identity and the reflection across $\mathscr{K}_0$. This identifies the second map of the top row of the main diagram as the twofold covering map from $\mathbb{RP}^3$ to $L(4, 1)$, with fiber the vertical elements of $\mathrm{isom}_f(K_0)$. We will identify $\mathrm{Isom}_v(K_0) \cap \mathrm{isom}_f(K_0)$ as the fiber of this covering map, by checking that it is C_2, generated by the isometry $f(1, i)$. By part (i) of Lemma 4.1, the elements of $\mathrm{isom}_f(K_0)$ are induced by the isometries $f(1, w_0)$ for $w_0 \in S^1$. Such an isometry is vertical precisely when $\overline{h}(1, w_0)$ acts as the identity or the antipodal map on U, since each fiber of K_0 is the image of the circles in S^3 which are the inverse images of antipodal points of U (since these are exactly the fibers of T_0 that are identified by elements of $G = F(D^*_{4m} \times \{1\})$). For $x_0 \in U$, we have $\overline{h}(1, w_0)(x_0) = \begin{pmatrix} \overline{w_0} & 0 \\ 0 & w_0 \end{pmatrix}(x_0) = \overline{w_0^2} x_0$. So $\overline{h}(1, w_0)$ is the identity or antipodal map of U exactly when $w_0 = \pm 1$ or $\pm i$. The cases $w_0 = \pm 1$ give $f(1, 1)$ and $f(1, -1)$, which are the identity on M since $F(-1, 1) = F(1, -1) \in G$.

Since $f(1, -1)$ is already in G, $f(1, i)$ and $f(1, -i)$ are the same isometry on K_0 and give the unique nonidentity element of $\mathrm{Isom}_v(K_0) \cap \mathrm{isom}_f(K_0)$.

Suppose now that $m = 1$. This time, both $\widetilde{\varrho}$ and ϱ are homeomorphisms, since $\mathscr{K}_0$ is just a geodesic arc connecting the two order-2 cone points of $\mathscr{O} = (S^2; 2, 2)$. From Tables 4.1 and 4.2, $\overline{h}$: $\mathrm{isom}(M) \to \mathrm{isom}(\mathscr{O})$ is just the projection from $S^1 \times S^1$ to its second coordinate. The first coordinate is left multiplication of S^3 by elements of S^1, which by part (ii) of Lemma 4.1 give exactly the elements of $\mathrm{isom}_f(K_0)$. Since $\overline{h}(x_0, 1)$ is the identity on S^2 for all these x_0, $\mathrm{isom}_f(K_0) = \mathrm{Isom}_v(K_0) \cap \mathrm{isom}_f(K_0)$. So the top row of the main diagram is simply the product fibration $S^1 \to S^1 \times S^1 \to S^1$, where the second map is projection to the second coordinate.

Finally, if both $m > 1$ and $n > 1$, the quotient orbifold is $(S^2; 2, 2, m)$ and as seen in the proof of Lemma 4.1, $\mathrm{isom}(\mathscr{K}_0)$ is a single point. Again part (ii) of Lemma 4.1 identifies $\mathrm{isom}_f(K_0, M)$ with the vertical isometries $\mathrm{Isom}_v(K_0)$ that are isotopic to the identity. So the top row of the main diagram is $S^1 \to S^1 \to \{1\}$.

In all three cases, the top row of the main diagram is a fibration. The proof will be completed by showing that the rightmost and leftmost vertical arrows of the main diagram are homotopy equivalences.

Suppose first that $n = 1$. We have a commutative diagram whose vertical maps are inclusions:

$$\begin{array}{ccccc}
\mathrm{Isom}(\mathscr{O}\,\mathrm{rel}\,\mathscr{K}_0) & \longrightarrow & \mathrm{isom}(\mathscr{O}) & \longrightarrow & \mathrm{isom}(\mathscr{K}_0, \mathscr{O}) \\
\downarrow & & \downarrow & & \downarrow \\
\mathrm{Diff}(\mathscr{O}\,\mathrm{rel}\,\mathscr{K}_0) & \longrightarrow & \mathrm{diff}(\mathscr{O}) & \longrightarrow & \mathrm{emb}(\mathscr{K}_0, \mathscr{O})
\end{array}$$

The bottom row is a fibration by Corollary 3.6, and we have already seen how to identify the top row with the covering fibration $C_2 \to \mathbb{RP}^3 \to L(4, 1)$. Each component of $\mathrm{Diff}(\mathscr{O}\,\mathrm{rel}\,\mathscr{K}_0)$ can be identified with $\mathrm{Diff}(D^2\,\mathrm{rel}\,\partial D^2)$, which is contractible by [64], so the left vertical arrow is a homotopy equivalence. The middle arrow is a homotopy equivalence by the main result of [19]. Consequently the right vertical arrow is a homotopy equivalence, which is also the right vertical arrow of the main diagram.

We have already seen that part (i) of Lemma 4.1 identifies S^1, the group of isometries of the form $f(1, w_0)$, with $\mathrm{isom}_f(K_0)$, so that $f(1, i)$ is the nontrivial element of $\mathrm{Isom}_v(K_0) \cap \mathrm{isom}_f(K_0)$. The group $\mathrm{Diff}_v(K_0) \cap \mathrm{diff}_f(K_0)$ consists of two contractible components, one in which the diffeomorphisms preserve the orientation of each fiber and the other in which they reverse it ($\mathrm{Diff}_v(K_0)$ consists of four contractible components, these two and two others represented by the same maps composed with a single Dehn twist about a vertical fiber). The identity map and $f(1, i)$ are points in these two components, so the left vertical arrow of the main diagram is also a homotopy equivalence.

A detailed analysis of $\mathrm{Diff}_v(K_0) \cap \mathrm{diff}_f(K_0)$ can proceed by regarding K_0 as a circle bundle over S^1, letting s_0 be a basepoint in S^1 and C be the fiber in K_0 which is the inverse image of s_0, and examining the commutative diagram

$$\begin{array}{ccccc}
\mathrm{Diff}_v(K_0 \,\mathrm{rel}\, C) \cap \mathrm{diff}_f(K_0) & \longrightarrow & \mathrm{Diff}_v(K_0) \cap \mathrm{diff}_f(K_0) & \longrightarrow & \mathrm{Diff}(C) \\
\downarrow & & \downarrow & & \downarrow \\
\mathrm{Diff}_f(K_0 \,\mathrm{rel}\, C) \cap \mathrm{diff}_f(K_0) & \longrightarrow & \mathrm{diff}_f(K_0) & \longrightarrow & \mathrm{emb}_f(C, K_0) \\
\downarrow & & \downarrow & & \downarrow \\
\mathrm{diff}(S^1 \,\mathrm{rel}\, s_0) & \longrightarrow & \mathrm{diff}(S^1) & \longrightarrow & \mathrm{emb}(s_0, S^1)
\end{array}$$

whose rows and columns are all fibrations (the first and middle rows using Theorem 3.6, the third row by the Palais–Cerf Restriction Theorem, the first and middle columns by Theorem 3.9, and the third column by Theorem 3.6). The spaces in this diagram are homotopy equivalent to the spaces shown here:

$$\begin{array}{ccccc}
\mathbb{Z} & \longrightarrow & C_2 \times \mathbb{R} & \longrightarrow & C_2 \times S^1 \\
\downarrow & & \downarrow & & \downarrow \\
\mathbb{Z} & \longrightarrow & S^1 \times \mathbb{R} & \longrightarrow & S^1 \times S^1 \\
\downarrow & & \downarrow & & \downarrow \\
1 & \longrightarrow & S^1 & \longrightarrow & S^1
\end{array}$$

When $n > 1$, the situation is quite a bit simpler. If $m = 1$, $\mathrm{emb}(\mathscr{K}_0, \mathscr{O})$ is just the embeddings of an arc in S^2 relative to two points, which is homotopy equivalent to $\mathrm{isom}(\mathscr{K}_0, \mathscr{O})$. For the left vertical arrow, $\mathrm{Diff}_v(\mathscr{K}_0) \cap \mathrm{diff}_f(K_0)$ has only one component, since a vertical diffeomorphism which reverses the direction of the fibers induces a nontrivial outer automorphism on $\pi_1(K_0)$. To see that $\mathrm{diff}_v(\mathscr{K}_0)$ is homotopy equivalent to a circle, we can fix a generic fiber C and a point c_0 in C, then lift a vertical diffeomorphism to a covering of $\mathscr{K}_0$ by $S^1 \times \mathbb{R}$ and equivariantly deform it to the isometry of $S^1 \times \mathbb{R}$ that has the same effect on a lift of c_0. This can be carried out canonically using the $\mathbb{R}$-coordinate, so actually gives a deformation retraction to $\mathrm{isom}_v(K_0)$. When $m > 1$, the situation is the same except that $\mathrm{isom}(\mathscr{K}_0, \mathscr{O})$ is a point and $\mathrm{emb}(\mathscr{K}_0, \mathscr{O})$ is contractible. □

4.5 Homotopy Type of the Space of Diffeomorphisms

We continue to use the notation of Sect. 4.4. The next theorem is our main technical result. It shows that parameterized families of embeddings of the base Klein bottle K_0 in M can be deformed to families of fiber-preserving embeddings.

Theorem 4.1. *If either $m > 1$ or $n > 1$, then the inclusion*

$$\mathrm{emb}_f(K_0, M) \to \mathrm{emb}(K_0, M)$$

is a homotopy equivalence.

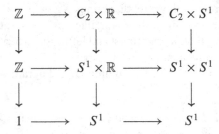

Its proof will be given in Sects. 4.6–4.8. From Theorem 4.1, we can deduce the Smale Conjecture for our three-manifolds for all cases except $M(1,1)$.

Theorem 4.2. *If $m > 1$ or $n > 1$, then the inclusion*

$$\mathrm{Isom}(M(m,n)) \to \mathrm{Diff}(M(m,n))$$

is a homotopy equivalence.

Proof (of Theorem 4.2 assuming Theorem 4.1). By Theorem 1.1, the inclusion is a bijection on path components, so we will restrict attention to the connected components of the identity map.

By Corollary 3.8, restriction of diffeomorphisms to embeddings defines a fibration

$$\mathrm{Diff}_f(M \text{ rel } K_0) \cap \mathrm{diff}_f(M) \to \mathrm{diff}_f(M) \to \mathrm{emb}_f(K_0, M).$$

Since any diffeomorphism in this fiber is orientation-preserving, it cannot locally interchange the sides of K_0. Therefore the fiber may be identified with a subspace consisting of path components of $\mathrm{Diff}_f(S^1 \times D^2 \text{ rel } S^1 \times \partial D^2)$. By Theorem 3.9, there is a fibration

$$\mathrm{Diff}_v(S^1 \times D^2 \text{ rel } S^1 \times \partial D^2) \to \mathrm{Diff}_f(S^1 \times D^2 \text{ rel } S^1 \times \partial D^2) \to \mathrm{Diff}(D^2 \text{ rel } \partial D^2)\,,$$

whose fiber is the group of vertical diffeomorphisms that take each fiber to itself. The base is contractible by [64], and it is not difficult to show that the fiber is contractible, so the restriction fibration becomes

$$\mathrm{diff}_f(M \text{ rel } K_0) \to \mathrm{diff}_f(M) \to \mathrm{emb}_f(K_0, M)$$

with contractible fiber. Similarly there is a fibration

$$\mathrm{diff}(M \text{ rel } K_0) \to \mathrm{diff}(M) \to \mathrm{emb}(K_0, M)\,.$$

The fact that it is a fibration is the Palais–Cerf Restriction Theorem, and the contractibility of the fiber uses [22]. We can now fit these into a diagram

$$\begin{array}{ccccc}
\mathrm{diff}_f(M \text{ rel } K_0) & \longrightarrow & \mathrm{diff}_f(M) & \longrightarrow & \mathrm{emb}_f(K_0, M) \\
\downarrow & & \downarrow & & \downarrow \\
\mathrm{diff}(M \text{ rel } K_0) & \longrightarrow & \mathrm{diff}(M) & \longrightarrow & \mathrm{emb}(K_0, M)\,.
\end{array}$$

The vertical maps are inclusions. By Theorem 4.1, the right hand vertical arrow is a homotopy equivalence. Since the fibers are both contractible, it follows that $\mathrm{diff}_f(M) \to \mathrm{emb}_f(K_0, M)$, $\mathrm{diff}(M) \to \mathrm{emb}(K_0, M)$, and $\mathrm{diff}_f(M) \to \mathrm{diff}(M)$ are homotopy equivalences.

The right-hand square of the previous diagram is the bottom square of the following diagram, whose vertical arrows are inclusions and whose horizontal arrows are obtained by restriction of maps to K_0:

$$\begin{array}{ccc} \mathrm{isom}(M) & \longrightarrow & \mathrm{isom}_f(K_0, M) \\ \downarrow & & \downarrow \\ \mathrm{diff}_f(M) & \longrightarrow & \mathrm{emb}_f(K_0, M) \\ \downarrow & & \downarrow \\ \mathrm{diff}(M) & \longrightarrow & \mathrm{emb}(K_0, M) \end{array}$$

Lemma 4.1 shows that $\mathrm{isom}(M) \to \mathrm{isom}_f(K_0, M)$ is a homeomorphism. Lemma 4.2 shows that $\mathrm{isom}_f(K_0, M) \to \mathrm{emb}_f(K_0, M)$ is a homotopy equivalence. Since the bottom square consists of homotopy equivalences, we conclude that $\mathrm{isom}(M) \to \mathrm{diff}_f(M)$ is a homotopy equivalence, hence so is the composite $\mathrm{isom}(M) \to \mathrm{diff}(M)$. □

4.6 Generic Position Configurations

Let S and T be smoothly embedded closed surfaces in a closed three-manifold M. A point x in $S \cap T$ is called a *singular* point if S is not transverse to T at x. There is a concept of finite multiplicity of such singular points, as described in Sect. 5 of [36] (another useful reference for these ideas is [8]). For a singular point x of finite multiplicity, either x is an isolated point of $S \cap T$, or $S \cap T$ meets a small disc neighborhood D^2 of x in T in a finite even number of smooth arcs running from x to ∂D^2, which are transverse intersections of S and T except at x (cf. Fig. 3, p. 1653 of [36]). Singular points are isolated on T, so by compactness $S \cap T$ will have only finitely many singular points.

We say that the surfaces are *in generic position* if all singular points of intersection are of finite multiplicity. Notice that $S \cap T$ is then a graph (with components that may be circles or isolated points) whose vertices are the singular points. Each vertex has even valence (possibly 0), since along each of the arcs of $S \cap T$ that emanates from the singular point, S crosses over to the (locally) other side of T.

Finite multiplicity intersections have the additional property that if $D^2 \times [-1, 1]$ is a product neighborhood of x which meets T in $D^2 \times \{0\}$, then for some $u_0 > 0$, S meets $D^2 \times \{u\}$ transversely for each u with $0 < |u| \leq u_0$ [36, Lemma 5.4]. Consequently, if T_u are the horizontal levels of a tubular neighborhood of T, with $u \in (-1, 0) \cup (0, 1)$ if T is two-sided, and $u \in (0, 1)$ if T is one-sided, then S is transverse to T_u for all u sufficiently close to 0.

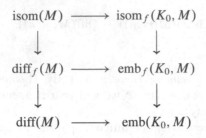

Now we specialize to the base Klein bottle $K_0 \subseteq M$, where as usual M denotes an $M(m, n)$ with either $m > 1$ or $n > 1$. To set notation, let T be the torus and fix a twofold covering from $T \times [-1, 1]$ to the twisted I-bundle neighborhood P of K_0, so that $T \times \{0\}$ is a twofold covering of K_0, and so that for $0 < u < 1$, the image T_u of $T \times \{u\}$ is a fibered torus. We call the T_u *levels*.

As usual, we write $\pi_1(K_0) = \langle a, b \mid bab^{-1} = a^{-1} \rangle$. For the meridional fibering, the fiber represents a, and for the longitudinal fibering, the exceptional fibers represent b, and the generic fiber b^2. We also recall from Sect. 4.4 that as a fibered submanifold of $M(m,n)$, K_0 has the meridional or longitudinal fibering according as $n = 1$ or $n > 1$.

Each T_u is the boundary of a tubular neighborhood P_u of K_0, and also bounds the solid torus $\overline{M - P_u}$, which we denote by R_u. For each $u > 0$, the elements a and b^2 generate the free abelian group $\pi_1(T_u)$, a subgroup of $\pi_1(P_u)$.

By a *meridian* in T_u we mean a simple loop in T_u which is essential in T_u but contractible in R_u. The meridians represent $(a^m b^{2n})^{\pm 1}$ in $\pi_1(T_u)$. By a *longitude* in T_u we mean a simple loop in T_u which represents a generator of the infinite cyclic group $\pi_1(R_u)$. The longitudes represent elements of $\pi_1(T_u)$ of the form $(a^p b^{2q} (a^m b^{2n})^k)^{\pm 1}$, where $pn - qm = \pm 1$, since these are precisely the elements whose intersection number with the meridians is ± 1. This leads us to the following observation.

Lemma 4.3. *Let ℓ be a loop in T_u which represents a or b^2 in $\pi_1(T_u)$. Then ℓ is not a meridian of R_u. If $(m,n) \neq (1,1)$, and ℓ is a longitude of R_u, then ℓ is isotopic in T_u to a fiber of the Seifert fibering of $M(m,n)$.*

Proof. Since neither of m nor n is 0, ℓ cannot be a meridian of R_u.

Suppose that ℓ represents a. If $n = 1$, then ℓ is a fiber of $M(m,n)$. If $n > 1$, then the longitudes are of the form $(a^p b^{2q} (a^m b^{2n})^k)^{\pm 1}$, where $pn - qm = \pm 1$. If a is a longitude, then $q + kn = 0$. But q and n are relatively prime, so this is impossible.

Suppose now that ℓ represents b^2. If $n > 1$, then ℓ is a fiber of $M(m,n)$. If $n = 1$, then the longitudes are of the form $(a(a^m b^2)^k)^{\pm 1}$. If b^2 is a longitude, then $1 + km = 0$. But when $n = 1$, we have $m > 1$, so this is impossible. □

The lemma fails for $M(1,1)$, for in that case an a-circle is a longitude of R_u which is not isotopic to a fiber of the longitudinal fibering, while a b^2-circle is a longitude not isotopic to a fiber of the meridional fibering.

If K is a Klein bottle in M that meets K_0 in generic position, then the intersection of K with the nearby levels is restricted by the next proposition, which is the main result of this section.

Proposition 4.1. *Suppose that $M = M(m,n)$ with $(m,n) \neq (1,1)$, and let K be a Klein bottle in M which is isotopic to K_0 and meets K_0 in generic position. Then there exists $u_0 > 0$ so that for each $u \leq u_0$, K is transverse to T_u, and each circle of $K \cap T_u$ is either inessential in T_u, or represents a or b^2 in $\pi_1(T_u)$.*

In order to prove Proposition 4.1, we introduce a special kind of isotopy. Suppose that L_0 is an embedded surface in a closed three-manifold N. A piecewise-linearly embedded surface S in N is said to be *flattened* (with respect to L_0) if it satisfies the following conditions:

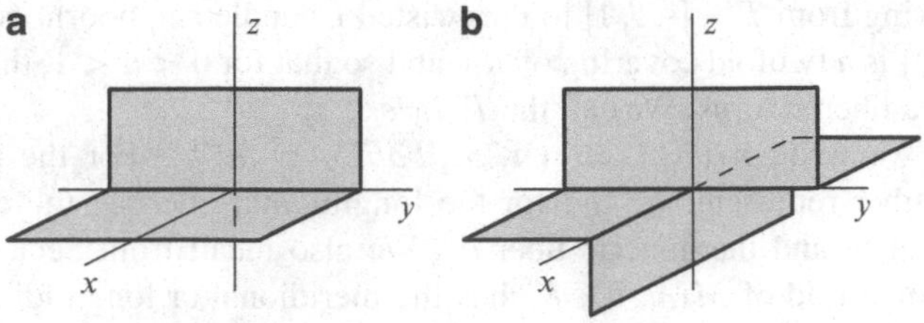

Fig. 4.1 Flattened surfaces, local picture

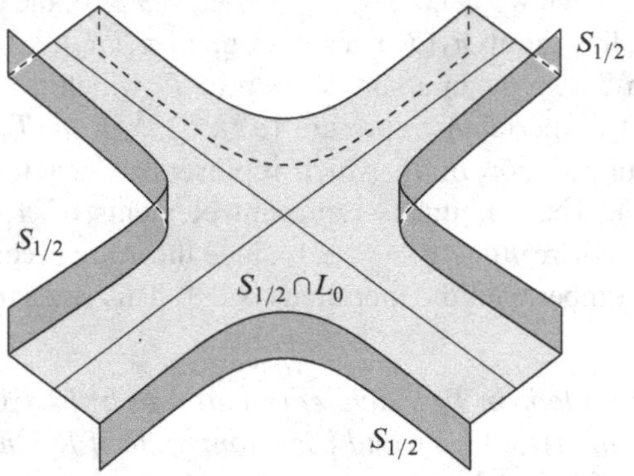

Fig. 4.2 A portion of the partially flattened surface $S_{1/2}$ near a point of $S_0 \cap L_0$ that was an ordinary saddle tangency. The intersecting *diagonal lines* are in $S_0 \cap L_0$. The horizontal surface is in $S_{1/2} \cap L_0$, while the *darker vertical strips* are in $S_{1/2}$ but not L_0

1. There is a 4-valent graph Γ (possibly with components which are circles) contained in L_0 such that $S \cap L_0$ consists of the closures of some of the connected components of $L_0 - \Gamma$.
2. Each point p in the interior of an edge of Γ has a neighborhood U for which the quadruple $(U, U \cap L_0, U \cap S, p)$ is PL homeomorphic to the configuration $(\mathbb{R}^3, \{(x, y, z) \mid z = 0\}, , \{(x, y, z) | \text{either } z = 0 \text{ and } x \geq 0, \text{ or } x = 0 \text{ and } z \geq 0\}, \{0\})$ (see Fig. 4.1a).
3. Each vertex v of Γ has a neighborhood U for which the quadruple $(U, U \cap L_0, U \cap S, v)$ is PL homeomorphic to the configuration $(\mathbb{R}^3, \{(x, y, z) \mid z = 0\}, \{(x, y, z) | \text{either } z = 0 \text{ and } xy \leq 0, \text{ or } x = 0 \text{ and } z \geq 0, \text{ or } y = 0 \text{ and } z \leq 0\}, \{0\})$ (see Fig. 4.1b).

In Fig. 4.1a, the graph Γ is the intersection of S with the y-axis. In Fig. 4.1b, Γ is the intersection of S with the union of the x- and y-axes, and in Fig. 4.2, it is the intersection of the horizontal portion in $S_{1/2} \cap L_0$ with the four vertical bands of $S_{1/2}$. The vertices of Γ are exactly the points that appear as the origin in a local picture as in Fig. 4.1b (or Fig. 4.3b below).

Lemma 4.4. *Let S_0 be a smoothly embedded surface in N which meets the one-sided surface L_0 in generic position. Denote by L_u the level surfaces in a tubular neighborhood $(L \times [-1, 1])/((x, u) \sim (\tau(x), -u))$ for some free involution τ of L*

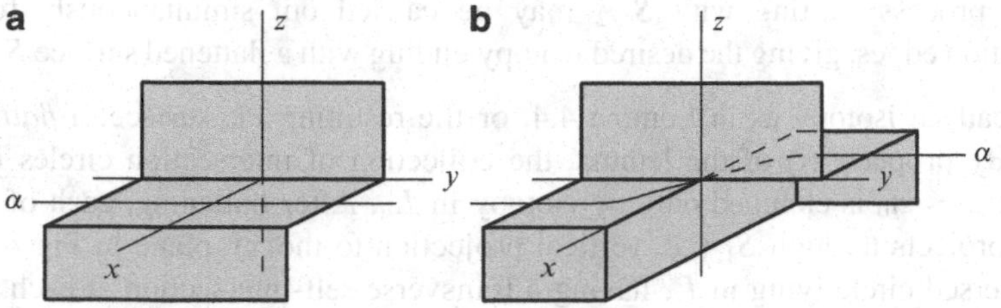

Fig. 4.3 Portions of a flattened surface near an original intersection arc α

with quotient L_0 (so $L_u = L_{-u}$). Then for some $u_0 > 0$, there is a PL isotopy S_t from S_0 to a PL embedded surface S_1 such that

(i) *Each S_t is transverse to L_u for $0 < u \leq u_0$.*
(ii) *S_1 is flattened with respect to L_0.*

Proof. Initially, S_0 meets L_0 in a graph, with tangencies at the vertices and transverse intersections on the open edges. We have already noted that there is a $u_0 > 0$ so that S_0 is transverse to L_u for all $0 < u \leq u_0$. The isotopy will only move S_0 in the region where $0 < u < u_0$. For $0 \leq t \leq 1$, we denote by S_t the image of S_0 at time t. With respect to u, points of S_t must move monotonically toward L_0, in such a way that the transversality required by condition (i) in the lemma is achieved.

Figure 4.2 illustrates the first portion of the isotopy, near a singular point x of $S_0 \cap L_0$. During the time $0 \leq t \leq 1/2$, a 2-disk neighborhood of x in S_0 moves onto a 2-disk neighborhood of x in L_0. In a neighborhood U of x, $S_0 \cap L_0$ consists of x together with a (possibly empty) collection of arcs $\alpha_1, \alpha_2, \ldots, \alpha_{2n}$ emanating from x, at which S_0 crosses alternately above and below L_0 as one travels around x on S_0. There is a neighborhood of x for which the angles of intersection of S_0 with L_0 are close to 0; the isotopy moves points only within such a neighborhood. At the end of the initial isotopy, there is a neighborhood U of x for which $S_{1/2} \cap L_0 \cap U$ is a regular neighborhood in L_0 of $\cup_{i=1}^{2n} \alpha_i$. Near interior points of the α_i, $S_{1/2}$ is positioned as in Fig. 4.3a, where L_0 is the horizontal plane and $S_{1/2}$ travels "up" on one side of α_i and "down" on the other. These isotopies may be performed simultaneously near each singular point of intersection.

The remainder of the isotopy will move points only in a small neighborhood of the original (open) edges of $S_0 \cap L_0$. Consider the closure α of such an edge. Initially, S_0 and L_0 were tangent at its endpoints (which may coincide), and nearly tangent near its endpoints, and $S_{1/2}$ actually coincides with L_0 at points of α near the endpoints. On the remainder, we continue to flatten $S_{1/2}$ so that it meets L_0 is a neighborhood of α, as shown locally in Fig. 4.3a.

When the two flattenings from the ends of α meet somewhere in the middle, it might happen that both go "up" on the same side of α, so that the flattening may be continued to achieve Fig. 4.3a at all points of the interior of α. It might happen, however, that one flattening goes "up" while the other goes "down" on the same side of α. In that case, we flatten to the local configuration in Fig. 4.3b, adding one such crossover point on each such edge α.

This process starting with $S_{1/2}$ may be carried out simultaneously for all intersection edges, giving the desired isotopy ending with a flattened surface S_1. □

We call an isotopy as in Lemma 4.4, or the resulting PL surface, a *flattening* of S_0. By property (i) of the lemma, the collection of intersection circles in L_u for $0 < u \leq u_0$ is changed only by isotopy in L_u. After flattening, each of these circles projects through S_1 (i.e. vertical projection to the xy-plane in Fig. 4.1) to an immersed circle lying in Γ, having a transverse self-intersection at each of its double points (which can occur only at vertices of Γ).

Proof (of Proposition 4.1). Suppose first that the intersection $K \cap K_0$ is transverse. Since K must meet every nearby level T_u transversely, it intersects P_u in Möbius bands and annuli, which after isotopy of K (keeping it transverse to level tori) may be assumed to be vertical in the I-bundle structure. The projection of T_u onto K_0 maps circles of $K \cap T_u$ onto circles of $K \cap K_0$ either homeomorphically or by twofold coverings. Only inessential and a- and b^2-circles can be inverse images of embedded circles in K_0. For suppose that a loop representing $a^k b^{2\ell}$ covers an embedded circle. Then it must have zero intersection number with its image under the covering transformation τ of T_u over K_0. Since a and b^2 have intersection number 1 in T_u, and $\tau(a) = a^{-1}$ and $\tau(b^2) = b^2$, the image represents $a^{-k} b^{2\ell}$ and the intersection number is $2k\ell$. Therefore the proposition holds when K meets K_0 transversely.

Suppose now that $K \cap K_0$ contains singular points. By Lemma 4.4, we can flatten K near K_0, without changing the isotopy classes in T_u of the loops $K \cap T_u$. After the flattening, $K \cap K_0$ consists of a valence 4 graph Γ, which is the image of the collection of disjoint simple closed curves $K \cap T_u$ under a twofold covering projection, together with some of the complementary regions of Γ in K_0, which we will call the *faces*. Each edge of Γ lies in the closure of exactly one face. It is convenient to choose an I-fibering of P_{u_0} so that $K \cap P_{u_0}$ lies in the union of $K \cap K_0$ and the I-fibers that meet Γ.

Suppose for contradiction that one of the circles in $K \cap T_u$ represents $a^k b^{2\ell}$ with $k\ell \neq 0$. Since K is geometrically incompressible (if not, then M would contain an embedded projective plane), there is an isotopy of K in M which eliminates the circles of $K \cap T_u$ that are inessential in T_u, without altering the remaining circles or destroying the flattened position of $K \cap P_u$.

At this point none of the components of Γ can be a circle. If so, then it would lie in a vertical annulus or Möbius band in $K \cap P_u$, and be the image of a 1 or twofold covering of a circle of $K \cap T_u$, but we have seen that only inessential, a-, and b^2-circles in T_u project along I-fibers to embedded circles in K_0. So we may assume that $K \cap T_u$ consists of disjoint circles each representing $a^k b^{2\ell}$. Since K is isotopic to K_0, each loop in T_u has zero mod 2-intersection number in M with K, and hence has even algebraic intersection number with $K \cap T_u$. Therefore $K \cap T_u$ consists of an even number of these circles; denote them by $A_1, A_2, \ldots, A_{2r}$.

The vertices of Γ are the images of the intersections of $\cup A_i$ with $\cup\tau(A_i)$ (note that by the properties of Γ, $\cup A_i$ and $\cup\tau(A_i)$ meet transversely). As above, we compute the intersection number to be

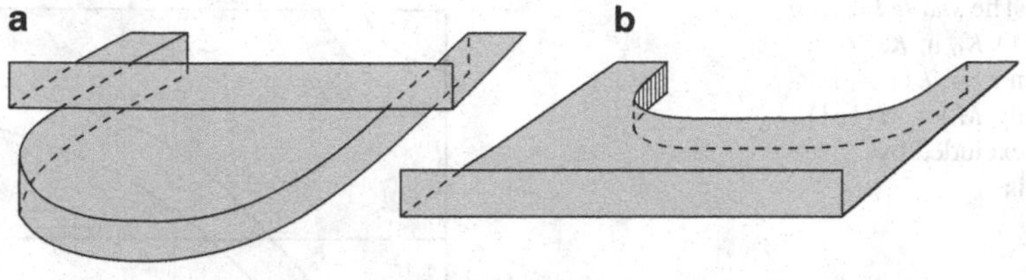

Fig. 4.4 Removal of a bigon by isotopy. The picture shows a portion of K near a bigon face of $K \cap K_0$, and K_0 is the horizontal plane containing the bigon. During the isotopy, the *top vertical* portion of K in (**a**) moves forward and the *bottom* one moves backward, ending with K in the position shown in (**b**). The bigon is eliminated from $K \cap K_0$, while the other two portions of intersection faces seen in (**a**) (which might be portions of the same face) are joined by a new horizontal band in $K \cap K_0$ seen in (**b**)

$$(\cup A_i) \cdot (\cup \tau(A_i)) = (2ra^k b^{2\ell}) \cdot (2ra^{-k} b^{2\ell}) = 4r^2\, 2k\ell\,.$$

Since $(\cup A_i) \cup (\cup \tau(A_i))$ is τ-invariant, each vertex of Γ is covered by two intersections, so Γ has at least $4r^2|k\ell|$ vertices.

We claim that each edge of Γ runs between two distinct vertices of Γ. Supposing to the contrary, we would see a crossing configuration as Fig. 4.1b, for which starting at the origin and traveling along one of the four edges of Γ that meet there returns to the origin along one of the other three edges without passing through another vertex. Suppose, for example, that the edge starts with the positive y-axis in Fig. 4.1b. Consider the right-hand orientation $(\vec{j}, -\vec{\imath}, \vec{k})$ at the origin in Fig. 4.1b. Travel out the edge e along the positive y-axis. On the edge, we can make a continuous choice of local orientation $(\vec{j}_t, -\vec{\imath}_t, \vec{k}_t)$ where $\vec{j}_t$ is a tangent vector to the edge, $-\vec{\imath}_t$ is the inward normal of $K \cap K_0$, and $\vec{k}_t$ is the inward normal of $\overline{K - K_0}$. Returning to the initial point of the edge, one approaches the origin along either the negative y-axis or the positive or negative x-axis, but on each of these axes the orientation $(\vec{j}_t, -\vec{\imath}_t, \vec{k}_t)$ is left-handed in Fig. 4.1b, contradicting the fact that M is orientable.

A similar argument shows that each face of $K \cap K_0$ has an even number of edges. Successive edges of a face meet at configurations as in Fig. 4.1b, and the orientations described in the previous paragraph change to the opposite orientation of M each time one passes to a new edge.

Consider a face that is a bigon. Since no edge has equal endpoints, the face must have two distinct vertices, as in Fig. 4.4a. The isotopy of K described in Fig. 4.4 eliminates this bigon and adds a band to $K \cap K_0$; this band is a (two-dimensional) 1-handle attached onto previous faces of $K \cap K_0$, and either combines two previous faces or is added onto a single previous face. Repeating, we move K by isotopy (not changing the isotopy classes of the loops of $K \cap T_u$) to eliminate all faces that are bigons. No component of Γ can be a circle, since as before this would force $K \cap T_u$ to have a component that is inessential or is an a- or b^2-curve. So each face of $K \cap K_0$ now contains at least four vertices.

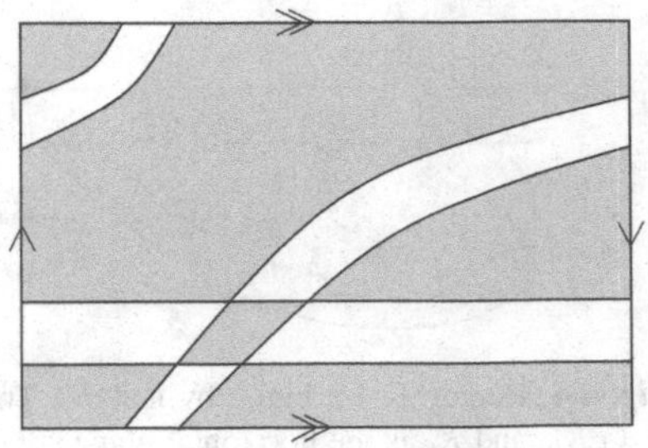

Fig. 4.5 The *shaded region* shows $K \cap K_0$ in K_0 for a case when $r = |k\ell| = 1$. Necessarily $M = M(1,1)$, which is excluded by hypothesis

The Euler characteristic of $K \cap P_u$ is at least $-2r$, since $\chi(K) = 0$ and $K \cap P_u$ has exactly $2r$ boundary components. Letting V, E, and F denote the number of vertices, edges, and faces of $K \cap K_0$, we have $E = 2V$ and $F \leq V/2$ (since each edge lies in exactly one face and each face has at least four edges). Therefore $-2r \leq \chi(K \cap P_u) = \chi(K \cap K_0) \leq -V/2$ (note that the latter estimate does not require that the faces themselves have Euler characteristic 1). Since $V \geq 4r^2|k\ell|$, it follows that $r|k\ell| \leq 1$, forcing $r = |k\ell| = 1$, $\chi(K \cap K_0) = -2$, $V = 4$, and $F = 2$. That is, $K \cap K_0$ consists of two faces, each a 4-gon, meeting at their four vertices. Since $|k\ell| = 1$, Γ is the image of two embedded circles of T_u each representing $a^{\pm 1}b^{\pm 2}$. Since $\chi(K \cap P_u) = \chi(K \cap K_0) = -2$ and $\chi(K) = 0$, each of these circles must bound a disk in R_u. This contradicts the hypothesis that $(m,n) \neq (1,1)$. □

Figure 4.5 shows $K \cap K_0$ for a Klein bottle K in $M(1,1)$ that *is* the flattening of a Klein bottle that meets every T_u close to K_0 in longitudes not homotopic to fibers, i.e. in loops representing ab^2.

4.7 Generic Position Families

In this section, we achieve the necessary generic position for a parameterized family. As usual, $M = M(m,n)$ with at least one of $m > 1$ or $n > 1$.

Proposition 4.2. *Let $F\colon D^k \to \mathrm{Emb}(K_0, M)$ be a parameterized family of embeddings of the standard Klein bottle K_0 into M. Then every open neighborhood of F in $C^\infty(D^k, \mathrm{Emb}(K_0, M))$ contains a map $G\colon D^k \to \mathrm{Emb}(K_0, M)$ for which $G(t)(K_0)$ is in generic position with respect to K_0 for all $t \in D^k$. Moreover, we may select G to be homotopic to F within the given neighborhood.*

Proof. From Lemma 5.2 of [36] (see also [8]), a G with each $G(t)(K_0)$ in generic position exists, and we need only verify that it may also be selected to be homotopic to F within the given neighborhood V.

Each $F(t)$ determines a bundle map from the restriction E of the tangent bundle of M to K_0 to the restriction $E(t)$ of the tangent bundle of M to $F(t)(K_0)$; in the directions tangent to K_0, it is the differential of $F(t)$, and it takes unit normals to unit normals. At each t, the Fréchet manifold of C^∞-sections of $E(t)$ whose image vectors all have length less than some sufficiently small ϵ corresponds, using the exponential map, to a neighborhood $W_\epsilon(t)$ of $F(t)$ in $\mathrm{Emb}(K_0, M)$. In particular, the zero section corresponds to $F(t)$. Since D^k is compact, we may fix a uniform value of ϵ for all $F(t)$.

An ϵ-small section $s(t)$, corresponding to an embedding $L(t)$, is isotopic to the zero section by sending each $s(x)$ to $(1-s)s(x)$. Via the exponential, this becomes a homotopy L_s with each $L_0(t) = L(t)$ and $L_1(t) = F(t)$, that is, a homotopy from L to F as elements of $C^\infty(D^k, \mathrm{Emb}(K_0, M))$. With respect to coordinates on $E(t)$, all partial derivatives of $s(t)$ move closer to zero s goes from 0 to 1, so those of L_s in M move closer to those of F. Consequently, provided that ϵ was small enough and G was selected so that each $G(t)$ was in $W_\epsilon(t) \cap V$, the G_s will remain in V. □

We are now ready for the main result of Sects. 4.6 and 4.7.

Theorem 4.3. *Let $F: D^k \to \mathrm{Emb}(K_0, M)$ be a parameterized family of Klein bottles in M. Assume that if $t \in \partial D^k$, then $F(t)$ is fiber-preserving and $F(t)(K_0) \neq K_0$. Then F is homotopic relative to ∂D^k to a family G such that for each $t \in D^k$, there exists $u > 0$ so that $G(t)(K_0)$ meets T_u transversely and each circle of $G(t)(K_0) \cap T_u$ is either inessential in T_u, or represents a or b^2 in $\pi_1(T_u)$.*

Proof. We first note that any embedded Klein bottle in M must meet K_0, since otherwise it would be embedded in the open solid torus $M - K_0$, so would admit an embedding into three-dimensional Euclidean space.

Recall that $M(m,n)$ is constructed from the I-bundle P over K_0 by attaching a solid torus. There is a u-coordinate on P, $0 \leq u \leq 1$, such that the points with $u = 0$ are K_0 and for each $0 < u \leq 1$, the points with u-coordinate equal to u are a "level" torus T_u. Fixing a parameter $t \in \partial D^k$, let $f: F(t)^{-1}(P - K_0) \to \mathrm{I}$ be the composition of $F(t)$ with projection to the u-coordinate of $P - K_0$. Since $F(t)(K_0)$ must meet K_0, and by hypothesis, $F(t)(K_0)$ does not equal K_0, the image of f contains an interval. By Sard's Theorem, almost all values of u are regular values of f, so there is some level T_u such that $F(t)(K_0)$ meets T_u transversely.

Since transversality is an open condition and ∂D^k is compact, there is a finite collection of open sets in D^k whose union contains ∂D^k and such that on each open set, there is a corresponding level T_u such that $F(t)(K_0)$ meets T_u transversely for every t in the open set. At points of ∂D^k, the intersection curves are fibers, so must be either a- or b^2-circles in $\pi_1(T_u)$. Choose a collar neighborhood $U = \partial D^k \times \mathrm{I}$ of ∂D^k, with $\partial D^k \times \{0\} = \partial D^k$, such that the closure $\overline{U}$ is contained in the union of these open sets. Since transversality is an open condition, there is an open neighborhood V of F in $C^\infty(D^k, C^\infty(K_0, M))$ such that for any map G in V and any $t \in U$, $G(t)$ is transverse to one of the corresponding levels, and $G(t)(K_0)$ intersects that level in loops representing either a or b^2.

Apply Proposition 4.2 to obtain a homotopy G'_s from F to a map G'_1 for which $G'_1(t)$ meets K_0 in generic position for every $t \in D^k$, and such that each G'_s lies in V. Define a new homotopy G_s that equals G'_s on $\overline{D^k - U}$ and carries out only the portion of G'_s from $s = 0$ to $s = r$ on each $\partial D^k \times \{r\} \subset U$. In particular, G_s is a homotopy relative to ∂D^k.

At each point t of U, $G_s(t)$ lies in V so is transverse to some level T_u and $G_s(K_0)$ intersects that level in loops representing either a or b^2. On $\overline{D^k - U}$, $G_s(K_0)$ meets K_0 in generic position, so by Proposition 4.1, it meets all T_u, for u sufficiently close to 0, transversely in loops which are either inessential in T_u or represent a or b^2 in $\pi_1(T_u)$. □

4.8 Parameterization

In this section we will complete the proof of Theorem 4.1. By definition, both $\mathrm{emb}(K_0, M)$ and $\mathrm{emb}_f(K_0, M)$ are connected, so $\pi_0(\mathrm{emb}(K_0, M), \mathrm{emb}_f(K_0, M)) = 0$. To prove that the higher relative homotopy groups vanish, we begin with a smooth map $F: D^k \to \mathrm{emb}(K_0, M)$, where $k \geq 1$, which takes all points of ∂D^k to $\mathrm{emb}_f(K_0, M)$. We will deform F, possibly changing the embeddings at parameters in ∂D^k but retaining the property that they are fiber-preserving, to a family which is fiber-preserving at every parameter. In fact, all deformations will be relative to ∂D^k, except for the first step.

Step 1: Obtain generic position

In order to apply Theorem 4.3, we must ensure that no $F_t(K_0)$ equals K_0 for $t \in \partial D^k$. Select a smooth isotopy J_s of M, $0 \leq s \leq 1$, with the following properties:

(a) J_0 is the identity of M.
(b) Each J_s is fiber-preserving.
(c) $J_1(K_0) \neq F(t)(K_0)$ for any $t \in \partial D^k$.

One construction of J_s is as follows. As elaborated in Sect. 4.4, the image of K_0 in the quotient orbifold $\mathcal{O}$ of the Hopf fibering is either a geodesic arc or a geodesic circle. Let A denote the image, let S be the endpoints of A if A is an arc and the empty set if A is a circle, and let T be the inverse image of S in M. Consider an isotopy j_t of $\mathcal{O}$, relative to S, that moves A to an arc or circle A' of large length.

By Theorem 3.9, the map $\mathrm{Diff}_f(M \text{ rel } T) \to \mathrm{Diff}(\mathcal{O} \text{ rel } S)$ induced by projection is a fibration, so j_t lifts to an isotopy J_t of M with J_0 the identity map of M. The image of $J_1(K_0)$ is A'. Since $F(t)$ is fiber-preserving for each $t \in \partial D^k$, its image is an arc or circle, and by compactness of ∂D^k, there is a maximum value for the lengths of these images. Provided that A' was selected to have length larger than this maximum, $J_1(K_0)$ cannot equal any $F(t)(K_0)$ for $t \in \partial D^k$.

We now perform a deformation F_s of F such that each $F_s(t) = J_s^{-1} \circ F(t)$. At each $t \in \partial D^k$, each $F_s(t)$ is fiber-preserving, and $F_1(t)(K_0) = J_1^{-1}(F(t)(K_0)) \neq K_0$. We can now apply Theorem 4.3 to further deform F relative to ∂D^k so that for each t, there is a value u, $0 < u \leq 1$, so that

(1) $F(t)$ is transverse to T_u.
(2) Every circle of $K_t \cap T_u$ is either inessential in T_u, or represents either a or b^2 in $\pi_1(T_u)$.

From now on, we will write K_t for $F(t)(K_0)$.

Step 2: Eliminate inessential intersection circles

The next step is to get rid of inessential intersections. Consider a single K_t and its associated level T_u. Notice first that each circle c of $K_t \cap T_u$ that bounds a (necessarily unique) 2-disk $D_T(c)$ in T_u also bounds a unique 2-disk $D_K(c)$ in K_t, since K_t is geometrically incompressible. We claim that if $D_K(c)$ is innermost among all such disks on K_t, then the interior of $D_K(c)$ is disjoint from T_u. If not, then there is a smaller disk E in $D_K(c)$ such that ∂E is essential in T_u and the interior of E is disjoint from T_u. Now E cannot be contained in P_u, since T_u is incompressible in P_u, so E must be a meridian disk of R_u. But then, ∂E is a circle of $K_t \cap T_u$ which is essential in T_u but does not represent either a or b^2, contradicting (2) and establishing the claim. We conclude that if $D_K(c)$ is innermost, then $D_K(c)$ and $D_T(c)$ bound a unique 3-ball $B(c)$ in M that meets T_u only in $D_T(c)$.

We now follow the procedure of Hatcher described in [23, 25] to deform the family F to eliminate the circles of $K_t \cap T_u$ that are inessential in T_u. It is not difficult to adapt the procedure to our situation, in fact a few simplifications occur, but since this is a crucial part of our argument and these methods are unfamiliar to many, we will navigate through the details. We will follow [25], as it is an easier read than [23], and its numbered formulas are convenient for referencing in our discussion. Start at the proof of the main theorem on p. 2 of [25]. Our K_t and T_u are in the role of the surfaces M_t and N_i in [25]. We ignore the points called p_t there, which are irrelevant for us (since a loop can bound at most one disk in K_t or T_u). Only notational substitutions are needed to obtain the conditions (1)–(3) and (5)–(6), (5_ϵ), and (6_ϵ) in [25] (condition (4) there concerns the irrelevant p_t), and the conditions called (a) and (b) there are assumed inductively as before. We have already seen that the disks $D_K(c_t)$ and $D_T(c_t)$ bound a unique 3-ball $B(c_t)$—this replaces the hypothesis of the main theorem of [25] that any two essential 2-spheres in M are isotopic. The argument that the boundary of $B(c_t)$ has a corner with angle less than π along c_t applies in our case, since T_u cannot be contained in the 3-ball $B(c_t)$.

The individual isotopies that make up the deformations called M_{tu} in [25] (so they would be called K_{tu} for us) are constructed as before. A crucial point in Hatcher's method is that if the isotopies pushing $D_K(c_t)$ and $D_K(c_t')$ across $B(c_t)$ and $B(c_t')$ overlap in time, then the balls $B(c_t)$ and $B(c_t')$ must be disjoint, ensuring that the isotopies have disjoint support and do not interfere with each other. The verification that such a $B(c_t)$ and $B(c_t')$ must be disjoint proceeds as in [25]: If the isotopies overlap in time, then

(i) $D_K(c_t)$ is disjoint from $D_K(c_t')$ (condition (5_ϵ)).
(ii) $D_T(c_t)$ and $D_T(c_t')$ lie in different levels T_u and $T_{u'}$ (condition (6_ϵ)).
(iii) $D_K(c_t)$ is disjoint from $T_{u'}$ and $D_K(c_t')$ is disjoint from T_u (condition (b)).

Conditions (i)–(iii) show that the boundaries of $B(c_t)$ and $B(c_t')$ are disjoint, so $B(c_t)$ and $B(c_t')$ are either disjoint or nested. But neither T_u nor $T_{u'}$ is contained in a 3-ball in M, so nesting would contradict (iii). The remainder of the proof is completed without significant modification.

At the end of this process, for each $t \in D^k$, there is a value $u > 0$ so that in place of (2) above we have

(2′) Every intersection circle of K_t with T_u represents either a or b^2 in $\pi_1(T_u)$.

Step 3: Make the intersection circles fibers

Since a and b^2 are nontrivial elements of $\pi_1(M)$, the circles of $K_t \cap T_u$ are essential in K_t as well, so each component of $K_t \cap R_u$ must be either an annulus or a Möbius band. In fact, Möbius bands cannot occur. For the center circle of such a Möbius band would have intersection number 1 with K_t and intersection number 0 with K_0, contradicting the fact that K_t is isotopic to K_0.

We will use the procedure of [23, 25], this time to pull the annuli $K_t \cap R_u$ out of the R_u. The details of adapting [23, 25] are not quite as straightforward as in Step 2. Again, the K_t and T_u are in the role of M_t and N_t respectively, and the points p_t are irrelevant. Setting notation, for each parameter t in a ball B_i in D^k, K_t is transverse to a level T_{u_i}, and $K_t \cap R_{u_i}$ is a collection of annuli. These annuli are in the role of the disks $D_M(c)$ of [25]. Denote by C_t^i the annuli of $K_t \cap R_{u_i}$ whose boundary circles are not isotopic in T_{u_i} to fibers. Notice that C_t^i is empty for parameters t in ∂D^k. By condition (2′) and Lemma 4.3, any circle of $K_t \cap R_u$ that is not isotopic in T_u to a fiber is also not a longitude of R_u. So each such annulus a_t is parallel across a region $W(a_t)$ in R_{u_i} to a uniquely determined annulus A_t in T_{u_i}.

Let C_t be the union of the C_t^i for which $t \in B_i$. Any two annuli of C_t are either nested or disjoint on K_t. Again we consider functions $\varphi_t \colon C_t \to (0, 1)$ such that $\varphi_t(a_t) < \varphi_t(a_t')$ whenever $a_t \subset a_t'$; this is the version of condition (5) needed for our case. For example, we may take $\varphi_t(a_t)$ to be the area in K_0 of the inverse image of a_t with respect to the embedding $K_0 \to K_t$, where the area of K_0 is normalized to 1. The transversality trick of [25] achieves condition (6) as before, conditions (5_ϵ) and (6_ϵ) are again true by compactness, and conditions (a) and (b) are assumed inductively.

The angles of the regions $W(a_t)$ along the circles $a_t \cap T_{u_i}$ are less than π, this time simply because a_t is contained in R_{u_i}.

Again, the key point in defining the isotopies that pull the a_t across the regions $W(a_t)$ and out of the R_{u_i} is that is two of the isotopies on such regions $W(a_t)$ and $W(a_t')$ overlap in time, then $W(a_t)$ and $W(a_t')$ must be disjoint. When they do overlap in time, we have

(i) a_t is disjoint from a_t' (condition (5_ϵ)).
(ii) A_t and A_t' lie in different levels T_u and $T_{u'}$ (condition (6_ϵ)).

We also have

(iii) a_t is disjoint from $T_{u'}$, and a_t' is disjoint from T_u.

To see this, choose notation so that $T_{u'} \subset R_u$. Then a_t' is disjoint from T_u since $a_t' \subset R_u$. If a_t were to meet $T_{u'}$, then there would be a circle of $a_t \cap T_{u'}$ that is parallel in $\overline{R_u - R_{u'}}$ to a circle of $a_t \cap T_u$. The latter is not a longitude of R_u, so the circle of $a_t \cap T_{u'}$ is not a longitude of $R_{u'}$. So a_t contains an annulus of $K_t \cap R_{u'}$ that is in C_t, contradicting condition (5_ϵ) (that is, such an annulus would already have been eliminated earlier in the isotopy).

Conditions (i) and (ii) show that the boundaries of $W(a_t)$ and $W(a_t')$ are disjoint, so $W(a_t)$ and $W(a_t')$ are either disjoint or nested. Suppose for contradiction that they are nested. Again we choose notation so that $R_{u'} \subset R_u$. Since a_t is disjoint from $T_{u'}$, and $a_t' \subset W(a_t)$, we must have $R_{u'} \subset W(a_t)$. It follows that $W(a_t)$ contains a loop that generates $\pi_1(R_{u'})$ and hence generates $\pi_1(R_u)$. But $W(a_t)$ is a regular neighborhood of the annulus a_t, and the boundary circles of a_t were not longitudes of R_u, so this is contradictory. The remaining steps of the argument require no non-obvious modifications.

At the end of this process, we have in addition to (1) that

(2″) Every circle of $K_t \cap T_u$ is isotopic in T_u to a fiber of the Seifert fibering on M.

Step 4: Establish lemmas needed for the final step

To complete the argument, we require two technical lemmas.

Lemma 4.5. *Let T be a torus with a fixed S^1-fibering, and let $C_n = \cup_{i=1}^n S_i$ be a union of n distinct fibers. Then* $\operatorname{emb}_f(C_n, T) \to \operatorname{emb}(C_n, T)$ *is a homotopy equivalence. The same holds for the Klein bottle with either the meridional fibering or the longitudinal singular fibering.*

Proof. First consider a surface F other than the 2-sphere, the disk, or the projective plane, with a base point x_0 in the interior of F and an embedding $S^1 \subset F$ with $x_0 \in S^1$ which does not bound a disk in F. In the next paragraph, we will sketch an argument using [19] that $\operatorname{emb}((S^1, x_0), (\operatorname{int}(F), x_0))$ has trivial homotopy groups. The approach is awkward and unnatural, but we have found no short, direct way to deduce this fact from [19] or other sources.

By the Palais–Cerf Restriction Theorem, there is a fibration

$$\operatorname{Diff}(F \operatorname{rel} S^1) \cap \operatorname{diff}(F, x_0) \to \operatorname{diff}(F, x_0) \to \operatorname{emb}((S^1, s_0), (\operatorname{int}(F), x_0)) .$$

Since F is not the 2-sphere, disk, or projective plane, Proposition 2 of [19] shows that $\operatorname{diff}(F, x_0)$ has the same homotopy groups as $\operatorname{diff}_1(F, x_0)$, the subgroup of diffeomorphisms that induce the identity on the tangent space at x_0, and by Theorem 2 of [19], the latter is contractible. So we have isomorphisms

$$\pi_{q+1}(\operatorname{emb}((S^1, s_0), (\operatorname{int}(F), x_0))) \cong \pi_q(\operatorname{Diff}(F \operatorname{rel} S^1))$$

for $q \geq 1$, and

$$\pi_1(\operatorname{emb}((S^1, s_0), (\operatorname{int}(F), x_0))) \cong \pi_0(\operatorname{Diff}(F \operatorname{rel} S^1) \cap \operatorname{diff}(F, x_0)) .$$

Proposition 6 of [19] shows that the components of $\mathrm{Diff}(F \operatorname{rel} S^1)$ are contractible, so it remains to see that only one component of $\mathrm{Diff}(F \operatorname{rel} S^1)$ is contained in $\mathrm{diff}(F, x_0)$. That is, if $h \in \mathrm{Diff}(F \operatorname{rel} S^1) \cap \mathrm{diff}(F, x_0)$, then h is isotopic to the identity relative to S^1. This is an exercise in surface theory, using Lemma 1.4.2 of [71].

We now start with the torus case of the lemma. Choose notation so that the S_i lie in cyclic order as one goes around T, and fix basepoints s_i in S_i for each i. Consider the diagram

$$\begin{array}{ccccc}
\mathrm{emb}_f(S_n, T \operatorname{rel} s_n) & \longrightarrow & \mathrm{emb}_f(S_n, T) & \longrightarrow & \mathrm{emb}(s_n, T) \\
\downarrow & & \downarrow & & \downarrow = \\
\mathrm{emb}(S_n, T \operatorname{rel} s_n) & \longrightarrow & \mathrm{emb}(S_n, T) & \longrightarrow & \mathrm{emb}(s_n, T)\,.
\end{array}$$

The first row is a fibration by Corollary 3.13 and the second by the Palais–Cerf Restriction Theorem. The fiber of the top row is homeomorphic to $\mathrm{Diff}_+(S_n \operatorname{rel} s_n)$, the group of orientation-preserving diffeomorphisms, which is contractible. We have already seen that the fiber of the second row is contractible. Therefore the middle vertical arrow is a homotopy equivalence. For $n = 1$, this completes the proof, so we assume that $n \geq 2$.

Let A be the annulus that results from cutting T along S_n, and let A_0 be the interior of A. Consider the diagram

$$\begin{array}{ccccc}
\mathrm{emb}_f(S_{n-1}, A_0 \operatorname{rel} s_{n-1}) & \longrightarrow & \mathrm{emb}_f(S_{n-1}, A_0) & \longrightarrow & \mathrm{emb}(s_{n-1}, A_0) \\
\downarrow & & \downarrow & & \downarrow = \\
\mathrm{emb}(S_{n-1}, A_0 \operatorname{rel} s_{n-1}) & \longrightarrow & \mathrm{emb}(S_{n-1}, A_0) & \longrightarrow & \mathrm{emb}(s_{n-1}, A_0)\,.
\end{array}$$

As in the previous diagram, the rows are fibrations. As before, the fibers are contractible, so the middle vertical arrow is a homotopy equivalence. Now consider the diagram

$$\begin{array}{ccccc}
\mathrm{emb}_f(C_{n-1}, A_0 \operatorname{rel} S_{n-1}) & \longrightarrow & \mathrm{emb}_f(C_{n-1}, A_0) & \longrightarrow & \mathrm{emb}_f(S_{n-1}, A_0) \\
\downarrow & & \downarrow & & \downarrow \\
\mathrm{emb}(C_{n-1}, A_0 \operatorname{rel} S_{n-1}) & \longrightarrow & \mathrm{emb}(C_{n-1}, A_0) & \longrightarrow & \mathrm{emb}(S_{n-1}, A_0)\,.
\end{array}$$

The top row is a fibration by Corollary 3.5, and the bottom by the Palais–Cerf Restriction Theorem. The right vertical arrow was shown to be a homotopy equivalence by the previous diagram. For $n = 2$, both fibers are points, so the middle vertical arrow is a homotopy equivalence. But $\mathrm{emb}_f(C_{n-1}, A_0 \operatorname{rel} S_{n-1})$ can be identified with $\mathrm{emb}_f(C_{n-2}, A_0 - S_{n-1})$, and similarly for the non-fiber-preserving spaces. So induction on n shows that the middle vertical arrow is a homotopy equivalence for any value of n.

To complete the proof, we use the diagram

$$\begin{array}{ccccc} \mathrm{emb}_f(C_n, T \,\mathrm{rel}\, S_n) & \longrightarrow & \mathrm{emb}_f(C_n, T) & \longrightarrow & \mathrm{emb}_f(S_n, T) \\ \downarrow & & \downarrow & & \downarrow \\ \mathrm{emb}(C_n, T \,\mathrm{rel}\, S_n) & \longrightarrow & \mathrm{emb}(C_n, T) & \longrightarrow & \mathrm{emb}(S_n, T)\,. \end{array}$$

The rows are fibrations, as in the previous diagram. The right-hand vertical arrow is the case $n = 1$, already proven, and the map between fibers can be identified with $\mathrm{emb}_f(C_{n-1}, A_0) \to \mathrm{emb}(C_{n-1}, A_0)$, which has been shown to be a homotopy equivalence for all n.

For the Klein bottle case, the proof is line-by-line the same in the case of the meridional fibering. For the longitudinal singular fibering, the only difference is that rather than an annulus A, the first cut along S_n produces either one or two Möbius bands. □

Lemma 4.6. *Let Σ be a compact three-manifold with nonempty boundary and having a fixed Seifert fibering, and let F be a vertical two-manifold properly embedded in Σ. Let $\mathrm{emb}_{\partial f}(F, \Sigma)$ be the connected component of the inclusion in the space of (proper) embeddings for which the image of ∂F is a union of fibers. Then $\mathrm{emb}_f(F, \Sigma) \to \mathrm{emb}_{\partial f}(F, \Sigma)$ is a homotopy equivalence.*

To prove Lemma 4.6, we need a preliminary result.

Lemma 4.7. *The following maps induced by restriction are fibrations.*

(i) $\mathrm{emb}(F, \Sigma) \to \mathrm{emb}(\partial F, \partial\Sigma)$
(ii) $\mathrm{emb}_{\partial f}(F, \Sigma) \to \mathrm{emb}_f(\partial F, \partial\Sigma)$
(iii) $\mathrm{emb}_f(F, \Sigma) \to \mathrm{emb}_f(\partial F, \partial\Sigma)$

Proof. Parts (i) and (iii) are cases of Corollaries 3.11 and 3.12. Part (ii) follows from part (i) since $\mathrm{emb}_{\partial f}(F, \Sigma)$ is the inverse image of $\mathrm{emb}_f(\partial F, \partial\Sigma)$ under the fibration of part (i). □

Proof (of Lemma 4.6). First we use the following fibration from Theorem 3.9:

$$\mathrm{Diff}_v(\Sigma \,\mathrm{rel}\, \partial\Sigma) \cap \mathrm{diff}_f(\Sigma \,\mathrm{rel}\, \partial\Sigma) \to \mathrm{diff}_f(\Sigma \,\mathrm{rel}\, \partial\Sigma) \to \mathrm{diff}(\mathcal{O} \,\mathrm{rel}\, \partial\mathcal{O})$$

where $\mathcal{O}$ is the quotient orbifold of Σ and as usual Diff_v indicates the diffeomorphisms that take each fiber to itself. The full orbifold diffeomorphism group of $\mathcal{O}$ can be identified with a subspace consisting of path components of the diffeomorphism group of the two-manifold B obtained by removing the cone points from $\mathcal{O}$ (the subspace for which the permutation of punctures respects the local groups at the cone points). Since ∂B is nonempty, $\mathrm{diff}(B \,\mathrm{rel}\, \partial B)$ and therefore $\mathrm{diff}(\mathcal{O} \,\mathrm{rel}\, \partial\mathcal{O})$ are contractible. Since $\pi_1(\mathrm{diff}(\mathcal{O} \,\mathrm{rel}\, \partial\mathcal{O}))$ is trivial, the exact sequence of the fibration shows that $\mathrm{Diff}_v(\Sigma \,\mathrm{rel}\, \partial\Sigma) \cap \mathrm{diff}_f(\Sigma \,\mathrm{rel}\, \partial\Sigma)$ is connected, so is equal to $\mathrm{diff}_v(\Sigma \,\mathrm{rel}\, \partial\Sigma)$. It is not difficult to see that each component of $\mathrm{Diff}_v(\Sigma \,\mathrm{rel}\, \partial\Sigma)$

is contractible (see Lemma 3.15 for a similar argument), so we conclude that $\mathrm{diff}_f(\Sigma \text{ rel } \partial\Sigma)$ is contractible.

Next, consider the diagram

$$\begin{array}{ccccc} \mathrm{Diff}_f(\Sigma \text{ rel } F \cup \partial\Sigma) \cap \mathrm{diff}_f(\Sigma \text{ rel } \partial\Sigma) & \longrightarrow & \mathrm{diff}_f(\Sigma \text{ rel } \partial\Sigma) & \longrightarrow & \mathrm{emb}_f(F, \Sigma \text{ rel } \partial F) \\ \downarrow & & \downarrow & & \downarrow \\ \mathrm{Diff}(\Sigma \text{ rel } F \cup \partial\Sigma) \cap \mathrm{diff}(\Sigma \text{ rel } \partial\Sigma) & \longrightarrow & \mathrm{diff}(\Sigma \text{ rel } \partial\Sigma) & \longrightarrow & \mathrm{emb}(F, \Sigma \text{ rel } \partial F) \end{array}$$

where the rows are fibrations by Corollaries 3.8 and 3.1. We have already shown that the components of $\mathrm{Diff}_f(\Sigma \text{ rel } \partial\Sigma)$ and (by cutting along F) the components of $\mathrm{Diff}_f(\Sigma \text{ rel } F \cup \partial\Sigma)$ are contractible. By [22] (which, as noted in [22], extends to Diff using [24]), the components of $\mathrm{Diff}(\Sigma \text{ rel } \partial\Sigma)$ and $\mathrm{Diff}(\Sigma \text{ rel } F \cup \partial\Sigma)$ are contractible. Therefore to show that $\mathrm{emb}_f(F, \Sigma \text{ rel } \partial F) \to \mathrm{emb}(F, \Sigma \text{ rel } \partial F)$ is a homotopy equivalence it is sufficient to show that $\pi_0(\mathrm{Diff}_f(\Sigma \text{ rel } F \cup \partial\Sigma) \cap \mathrm{diff}_f(\Sigma \text{ rel } \partial\Sigma)) \to \pi_0(\mathrm{Diff}(\Sigma \text{ rel } F \cup \partial\Sigma) \cap \mathrm{diff}(\Sigma \text{ rel } \partial\Sigma))$ is bijective. It is surjective because every diffeomorphism of a Seifert-fibered three-manifold which is fiber-preserving on the (non-empty) boundary is isotopic relative to the boundary to a fiber-preserving diffeomorphism (Lemma VI.19 of Jaco [37]). It is injective because fiber-preserving diffeomorphisms that are isotopic are isotopic through fiber-preserving diffeomorphisms (see [71]).

The proof is completed by the following diagram in which the rows are fibrations by parts (iii) and (ii) of Lemma 4.7, and we have verified that the left vertical arrow is a homotopy equivalence.

$$\begin{array}{ccccc} \mathrm{emb}_f(F, \Sigma \text{ rel } \partial F) & \longrightarrow & \mathrm{emb}_f(F, \Sigma) & \longrightarrow & \mathrm{emb}_f(\partial F, \partial\Sigma) \\ \downarrow & & \downarrow & & \downarrow = \\ \mathrm{emb}(F, \Sigma \text{ rel } \partial F) & \longrightarrow & \mathrm{emb}_{\partial f}(F, \Sigma) & \longrightarrow & \mathrm{emb}_f(\partial F, \partial\Sigma) \end{array}$$

□

Step 5: Complete the proof

We can now complete the proof of Theorem 4.1 by deforming the family F to a fiber-preserving family. Since (1) and (2″) are open conditions, we can cover D^k by convex k-cells B_j, $1 \le j \le r$, having corresponding levels T_{u_j} for which (1) and (2″) hold throughout B_j. Also, we may slightly change the u-values, if necessary, to assume that the u_i are distinct. It is convenient to rename the B_j so that $u_1 < u_2 < \cdots < u_r$, that is, so that the levels T_{u_j} sit farther away from K_0 as j increases.

Choose a PL triangulation Δ of D^k sufficiently fine so that each i-cell lies in at least one of the B_j. The deformation of F will take place sequentially over the i-skeleta of Δ. It will never be necessary to change F at points of ∂D^k.

Suppose first that τ is a 0-simplex of Δ. Let $j_1 < j_2 < \cdots < j_s$ be the values of j for which $\tau \subseteq B_j$. By condition (2″), each intersection circle of K_τ with each

T_{j_q} is isotopic in T_{j_q} to a fiber of the Seifert fibering. We claim that they are also isotopic on K_τ to an image of a fiber of K_0 under $F(\tau)$. Since K_τ is isotopic to K_0 and the intersection circles are two-sided in K_τ, each intersection circle is isotopic in M to an a-loop or a b^2-loop in K_0. When $m = 1$, b^2 is the generic fiber of M, and a is not isotopic in M to b^2 since $a = b^{2n}$ and $n \neq 1$. When $n = 1$, a is the fiber of M, and b^2 is not isotopic to a since $a^m = b^2$ and $m > 1$. So the isotopy from K_τ to K_0 carries the intersection loops to loops in K_0 representing the fiber. But a-loops are nonseparating and b^2-loops are separating, so the intersection loops must be isotopic in K_τ to the image of the fiber of K_0 under $F(\tau)$.

We may deform the parameterized family near τ, retaining transverse intersection with each T_{u_j} for which $\tau \in B_j$, so that the intersection circles of K_τ with these T_{u_j} are fibers and images of fibers. To accomplish this, first change $F(\tau)$ by an ambient isotopy of M that preserves levels and moves the intersection circles onto fibers in the T_{u_j}. Now, consider the inverse images of these circles in K_0. We have seen that there is an isotopy that moves them to be fibers, changing $F(\tau)$ by this isotopy (and tapering it off in a small neighborhood of τ in D^k) we may assume that the intersection circles are fibers of K_τ as well. Now, using Lemma 4.6 successively on the solid torus R_{u_s}, the product regions $\overline{R_{u_{j-1}} - R_{u_j}}$ for $j = j_s, j_{s-1}, \ldots, j_2$, and the twisted I-bundle $P_{u_{j_1}}$, deform $F(\tau)$ to be fiber-preserving. These isotopies preserve the levels T_{u_j} for which $\tau \in B_j$, so may be tapered off near τ so as not to alter any other transversality conditions.

Inductively, suppose that $F(t)$ is fiber-preserving for each t lying in any i-simplex of Δ. Let τ be an $(i+1)$-simplex of Δ. For each $t \in \partial\tau$, $F(t)$ is fiber-preserving. Consider a level T_{u_j} for which $\tau \subset B_j$. For each $t \in \tau$, the restriction of $F(t)$ to the inverse image of T_{u_j} is a parameterized family of embeddings of a family of circles into T_{u_j}, which embeds to fibers at each $t \in \partial\tau$. By Lemma 4.5, there is a deformation of $F|_\tau$, relative to $\partial\tau$, which makes each $K_t \cap T_{u_j}$ consist of fibers in T_{u_j}. We may select the deformation so as to move image points of each $F(t)$ only very near T_{u_j}, and thereby not alter transversality with any other T_{u_ℓ}. Now, the restriction of the $F(t)^{-1}$ to the intersection circles is a family of embeddings of a collection of circles into K_0, which are fibers at points in $\partial\tau$. Using Lemma 4.5 we may alter $F|_\tau$, relative to $\partial\tau$ and without changing the images $F(t)(K_0)$, so that the intersection circles are fibers of K_0 as well. We repeat this for all ℓ such that $\tau \subset B_\ell$. Using Lemma 4.6 as before, proceeding from $R_{u_{j_s}}$ to $P_{u_{j_1}}$, deform F on τ, relative to $\partial\tau$, to be fiber-preserving for all parameters in τ. This completes the induction step and the proof of Theorem 4.1.

Chapter 5
Lens Spaces

Recall that we always use the term *lens space* to mean a three-dimensional lens space $L(m,q)$ with $m \geq 3$. In addition, we always select q so that $1 \leq q < m/2$.

In this chapter, we will prove Theorem 1.3, the Smale Conjecture for Lens Spaces. The argument is regrettably quite lengthy. It uses a lot of combinatorial topology, but draws as well on some mathematics unfamiliar to many low-dimensional topologists. We have already seen some of that material in earlier chapters, but we will also have to use the Rubinstein–Scharlemann method, reviewed in Sect. 5.6, and some results from singularity theory, presented in Sect. 5.8.

The next section is a comprehensive outline of the entire proof. We hope that it will motivate the various technical complications that ensue.

5.1 Outline of the Proof

Some initial reductions, detailed in Sect. 5.2, reduce the Smale Conjecture for Lens Spaces to showing that the inclusion $\mathrm{diff}_f(L) \to \mathrm{diff}(L)$ is an isomorphism on homotopy groups. Here, $\mathrm{diff}(L)$ is the connected component of the identity in $\mathrm{Diff}(L)$, and $\mathrm{diff}_f(L)$ is the connected component of the identity in the group of diffeomorphisms that are fiber-preserving with respect to a Seifert fibering of L induced from the Hopf fibering of its universal cover, S^3. To simplify the exposition, most of the argument is devoted just to proving that $\mathrm{diff}_f(L) \to \mathrm{diff}(L)$ is surjective on homotopy groups, that is, that a map from S^d to $\mathrm{diff}(L)$ is homotopic to a map into $\mathrm{diff}_f(L)$. The injectivity is obtained in Sect. 5.13 by a combination of tricks and minor adaptations of the main program.

Of course, a major difficulty in working with elliptic three-manifolds is their lack of incompressible surfaces. In their place, we use another structure which has a certain degree of essentiality, called a *sweepout*. This means a structure on L as a quotient of $P \times \mathrm{I}$, where P is a torus, in which $P \times \{0\}$ and $P \times \{1\}$ are collapsed to core circles of the solid tori of a genus 1 Heegaard splitting of L. For $0 < u \leq 1$,

S. Hong et al., *Diffeomorphisms of Elliptic 3-Manifolds*, Lecture Notes in Mathematics 2055, DOI 10.1007/978-3-642-31564-0_5,

$P \times \{t\}$ becomes a Heegaard torus in L, denoted by P_u, and called a *level*. The sweepout is chosen so that each P_u is a union of fibers. Sweepouts are examined in Sect. 5.5.

Start with a parameterized family of diffeomorphisms $f: L \times S^d \to L$, and for $u \in S^d$ denote by f_u the restriction of f to $L \times \{u\}$. The procedure that deforms f to make each f_u fiber-preserving has three major steps.

Step 1 ("finding good levels") is to perturb f so that for each u, there is some pair (s, t) so that $f_u(P_s)$ intersects P_t transversely, in a collection of circles each of which is either essential in both $f_u(P_s)$ and P_t (a *biessential* intersection), or inessential in both (a *discal* intersection), and at least one intersection circle is biessential. This pair is said to intersect in *good position,* and if none of the intersections is discal, in *very good position.* These concepts are developed in Sect. 5.4, after a preliminary examination of annuli in solid tori in Sect. 5.3.

To accomplish Step 1, the methodology of Rubinstein and Scharlemann in [58] is adapted. This is reviewed in Sect. 5.6. First, one perturbs f to be in "general position," as defined in Sect. 5.8. The intersections of the $f_u(P_s)$ and P_t are then sufficiently well-controlled to define a *graphic* in the square I^2. That is, the pairs (s, t) for which $f_u(P_s)$ and P_t do not intersect transversely form a graph embedded in the square. The complementary regions of this graph in I^2 are labeled according to a procedure in [58], and in Sect. 5.9 we show that the properties of general position salvage enough of the combinatorics of these labels developed in [58] to deduce that at least one of the complementary regions consists of pairs in good position.

Perhaps the hardest work of the proof, and certainly the part that takes us furthest from the usual confines of low-dimensional topology, is the verification that sufficient "general position" can be achieved. Since we use parameterized families, we must allow $f_u(P_s)$ and P_t to have large numbers of tangencies, some of which may be of high order. It turns out that to make the combinatorics of [58] go through, we must achieve that at each parameter *there are at most finitely many pairs* (s, t) *where* $f_u(P_s)$ *and* P_t *have multiple or high-order tangencies* (at least, for pairs not extremely close to the boundary of the square). To achieve the necessary degree of general position, we use results of a number of people, notably Bruce [8] and Sergeraert [63].

The need for this kind of general position is indicated in Sect. 5.7, where we construct a pair of sweepouts of $S^2 \times S^1$ with all tangencies of Morse type, but having no pair of levels intersecting in good position. Although we have not constructed a similar example for an $L(m, q)$, we see no reason why one could not exist.

Step 2 ("from good to very good") is to deform f to eliminate the discal intersections of $f_u(P_s)$ and P_t, for certain pairs in good position that have been found in Step 1, so that they intersect in very good position. This is an application of Hatcher's parameterization methods [22]. One must be careful here, since an isotopy that eliminates a discal intersection can also eliminate a biessential intersection, and if all biessential intersections were eliminated by the procedure, the resulting pair would no longer be in very good position. Lemma 5.11 ensures that not all biessential intersections will be eliminated.

Step 3 ("from very good to fiber-preserving") is to use the pairs in very good position to deform the family so that each f_u is fiber-preserving. This is carried out in Sects. 5.11 and 5.12. The basic idea is first to use the biessential intersections to deform the f_u so that $f_u(P_s)$ actually equals P_t (for certain (s,t) pairs that originally intersected in good position), then use known results about the diffeomorphism groups of surfaces and Haken three-manifolds to make the f_u fiber-preserving on P_s and then on its complementary solid tori. This process is technically complicated for two reasons. First, although a biessential intersection is essential in both tori, it can be contractible in one of the complementary solid tori of P_t, and $f_u(P_s)$ can meet that complementary solid torus in annuli that are not parallel into P_t. So one may be able to push the annuli out from only one side of P_t. Secondly, the fitting together of these isotopies requires one to work with not just one level but many levels at a single parameter.

Two natural questions are whether Bonahon's original method for determining the mapping class group $\pi_0(\mathrm{Diff}(L))$ [6] can be adapted to the parameterized setting, and whether our methodology can be used to recover his results. Concerning the first question, we have had no success with this approach, as we see no way to perturb the family to the point where the method can be started at each parameter. For the second, the answer is yes. In fact, the key geometric step of [6] is the isotopy uniqueness of genus-one Heegaard surfaces in L, which was already reproven in Rubinstein and Scharlemann's original work [58, Corollary 6.3].

5.2 Reductions

In this section, we carry out some initial reductions. The Conjecture will be reduced to a purely topological problem of deforming parameterized families of diffeomorphisms to families of diffeomorphisms that preserve a certain Seifert fibering of L.

By Theorem 1.1, it is sufficient to prove that $\mathrm{isom}(L) \to \mathrm{diff}(L)$ is a homotopy equivalence. And we have seen that this follows once we prove that $\mathrm{isom}(L) \to \mathrm{diff}(L)$ is a homotopy equivalence.

Section 1.4 of [46] gives a certain way to embed $\pi_1(L)$ into SO(4) so that its action on S^3 is fiber-preserving for the fibers of the Hopf bundle structure of S^3. Consequently, this bundle structure descends to a Seifert fibering of L, which we call the *Hopf fibering* of L. If $q = 1$, this Hopf fibering is actually an S^1-bundle structure, while if $q > 1$, it has two exceptional fibers with invariants of the form (k, q_1), (k, q_2) where $k = m/\gcd(q-1, m)$ (see Table 4 of [46]). We will always use the Hopf fibering as the Seifert-fibered structure of L.

Theorem 2.1 of [46] shows that (since $m > 2$) every orientation-preserving isometry of L preserves the Hopf fibering on L. In particular, $\mathrm{isom}(L) \subset \mathrm{diff}_f(L)$, so there are inclusions

$$\mathrm{isom}(L) \to \mathrm{diff}_f(L) \to \mathrm{diff}(L)\,.$$

Theorem 5.1. *The inclusion* $\mathrm{isom}(L) \to \mathrm{diff}_f(L)$ *is a homotopy equivalence.*

Proof. The argument is similar to the latter part of the proof of Theorem 4.2, so we only give a sketch. There is a diagram

$$\begin{array}{ccccc} S^1 & \longrightarrow & \mathrm{isom}(L) & \longrightarrow & \mathrm{isom}(L_0) \\ \downarrow & & \downarrow & & \downarrow \\ \mathrm{vert}(L) & \longrightarrow & \mathrm{diff}_f(L) & \longrightarrow & \mathrm{diff}_{orb}(L_0) \end{array}$$

where L_0 is the quotient orbifold and $\mathrm{diff}_{orb}(L_0)$ is the group of orbifold diffeomorphisms of L_0, and $\mathrm{vert}(L)$ is the group of vertical diffeomorphisms. The first row is a fibration, in fact an S^1-bundle, and the second row is a fibration by Theorem 3.9. The vertical arrows are inclusions. When $q = 1$, L_0 is the 2-sphere and the right-hand vertical arrow is the inclusion of SO(3) into $\mathrm{diff}(S^2)$, which is a homotopy equivalence by [64]. When $q > 1$, L_0 is a 2-sphere with two cone points, $\mathrm{isom}(L_0)$ is homeomorphic to S^1, and $\mathrm{diff}_{orb}(L_0)$ is essentially the connected component of the identity in the diffeomorphism group of the annulus. Again the right-hand vertical arrow is a homotopy equivalence. The left-hand vertical arrow is a homotopy equivalence in both cases, so the middle arrow is as well. □

Theorem 5.1 reduces the Smale Conjecture for Lens Spaces to proving that the inclusion $\mathrm{diff}_f(L) \to \mathrm{diff}(L)$ is a homotopy equivalence. For this it is sufficient to prove that for all $d \geq 1$, any map $f\colon (D^d, S^{d-1}) \to (\mathrm{diff}(L), \mathrm{diff}_f(L))$ is homotopic, through maps taking S^{d-1} to $\mathrm{diff}_f(L)$, to a map from D^d into $\mathrm{diff}_f(L)$. To simplify the exposition, we work until the final section with a map $f\colon S^d \to \mathrm{diff}(L)$ and show that it is homotopic to a map into $\mathrm{diff}_f(L)$. In the final section, we give a trick that enables the entire procedure to be adapted to maps $f\colon (D^d, S^{d-1}) \to (\mathrm{diff}(L), \mathrm{diff}_f(L))$, completing the proof.

5.3 Annuli in Solid Tori

Annuli in solid tori will appear frequently in our work. Incompressible annuli present little difficulty, but we will also need to examine compressible annuli, whose behavior is more complicated. In this section, we provide some basic definitions and lemmas.

A loop α in a solid torus V is called a *longitude* if its homotopy class is a generator of the infinite cyclic group $\pi_1(V)$. If in addition there is a product structure $V = S^1 \times D^2$ for which $\alpha = S^1 \times \{0\}$, then α is called a *core circle* of V. A subset of a solid torus V is called a *core region* when it contains a core circle of V. An embedded circle in ∂V which is essential in ∂V and contractible in V is called a *meridian* of V; a properly embedded disk in V whose boundary is a meridian is called a *meridian disk* of V.

Annuli in solid tori will always be assumed to be properly embedded, which for us includes the property of being transverse to the boundary, unless they are actually contained in the boundary. The next three results are elementary topological facts, and we do not include proofs.

Proposition 5.1. *Let A be a boundary-parallel annulus in a solid torus V, which separates V into V_0 and V_1, and for $i = 0, 1$, let $A_i = V_i \cap \partial V$. Then A is parallel to A_i if and only if V_{1-i} is a core region.*

Proposition 5.2. *Let A be a properly embedded annulus in a solid torus V, which separates V into V_0 and V_1, and let $A_i = V_i \cap \partial V$. The following are equivalent:*

1. *A contains a longitude of V.*
2. *A contains a core circle of V.*
3. *A is parallel to both A_0 and A_1.*
4. *Both V_0 and V_1 contain longitudes of V.*
5. *Both V_0 and V_1 are core regions of V.*

An annulus satisfying the conditions in Proposition 5.2 is said to be *longitudinal.* A longitudinal annulus must be incompressible.

Proposition 5.3. *Let V be a solid torus and let $\cup A_i$ be a union of disjoint boundary-parallel annuli in V. Let C be a core circle of V that is disjoint from $\cup A_i$. For each A_i, let V_i be the closure of the complementary component of A_i that does not contain C, and let $B_i = V_i \cap \partial V$. Then A_i is parallel to B_i. Furthermore, either*

1. *No A_i is longitudinal, and exactly one component of $V - \cup A_i$ is a core region, or*
2. *Every A_i is longitudinal, and every component of $V - \cup A_i$ is a core region.*

There are various kinds of compressible annuli in solid tori. For example, there are boundaries of regular neighborhoods of properly embedded arcs, possibly knotted. Also, there are annuli with one boundary circle a meridian and the other a contractible circle in the boundary torus. When both boundary circles are meridians, we call the annulus *meridional.* As shown in Fig. 5.1, meridional annuli are not necessarily boundary-parallel.

Although meridional annuli need not be boundary-parallel, they behave homologically as though they were, and as a consequence any family of meridional annuli misses some longitude.

Lemma 5.1. *Let $A_1, \ldots, A_n$ be disjoint meridional annuli in a solid torus V. Then:*

1. *Each A_i separates V into two components, $V_{i,0}$ and $V_{i,1}$, for which A_i is incompressible in $V_{i,0}$ and compressible in $V_{i,1}$.*
2. *$V_{i,1}$ contains a meridian disk of V.*
3. *$\pi_1(V_{i,0}) \to \pi_1(V)$ is the zero homomorphism.*
4. *The intersection of the $V_{i,1}$ is the unique component of the complement of $\cup A_i$ that contains a longitude of V.*

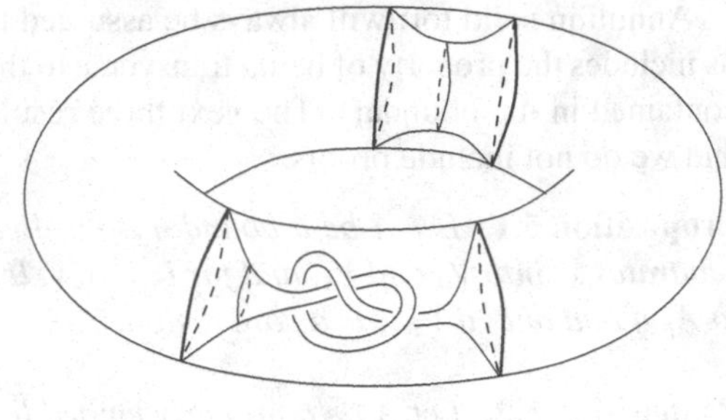

Fig. 5.1 Meridional annuli in a solid torus

Proof. For each i, every loop in V has even algebraic intersection with A_i, since every loop in ∂V does, so A_i separates V. Since A_i is not incompressible, it must be compressible in one of its complementary components, $V_{i,1}$, and since V is irreducible, A_i must be incompressible in the other complementary component, $V_{i,0}$.

Notice that $V_{i,1}$ must contain a meridian disk of V. Indeed, if K is the union of A_i with a compressing disk in $V_{i,1}$, then two of the components of the frontier of a regular neighborhood of K in V are meridian disks of $V_{i,1}$. Consequently, $\pi_1(V_{i,0}) \to \pi_1(V)$ is the zero homomorphism. The Mayer–Vietoris sequence shows that $H_1(A_i) \to H_1(V_{i,0})$ and $H_1(V_{i,1}) \to H_1(V)$ are isomorphisms.

Let V_1 be the intersection of the $V_{i,1}$, and let V_0 be the union of the $V_{i,0}$. The Mayer–Vietoris sequence shows that V_1 is connected, and that $H_1(V_1) \to H_1(V)$ is an isomorphism, so V_1 contains a longitude of V. For any i, j, either $V_{i,1} \subseteq V_{j,1}$ or $V_{j,1} \subseteq V_{i,1}$, since otherwise $H_1(V_{i,1}) \to H_1(V_{j,0}) \to H_1(V)$ would be the zero homomorphism. Therefore the intersection $V_1 = \cap V_{i,1}$ is equal to some $V_{k,1}$, and in particular it contains a longitude of V. No other complementary component of $\cup A_i$ contains a longitude, since each such component lies in $V_{k,0}$, all of whose loops are contractible in V. □

5.4 Heegaard Tori in Very Good Position

A *Heegaard torus* in a lens space L is a torus that separates L into two solid tori. In this section we will develop some properties of Heegaard tori. Also, we introduce the concepts of discal and biessential intersection circles, good position, and very good position, which will be used extensively in later sections.

When P is a Heegaard torus bounding solid tori V and W, and Q is a Heegaard torus contained in the interior of V, Q need not be parallel to ∂V. For example, start with a core circle in V, move a small portion of it to ∂V, then pass it across a meridian disk of W and back into V. This moves the core circle to its band-connected sum in V with an (m, q)-curve in ∂V. By varying the choice of band—for example, by twisting it or tying knots in it—and by iterating this construction, one can construct complicated knotted circles in V which are isotopic in L to a core

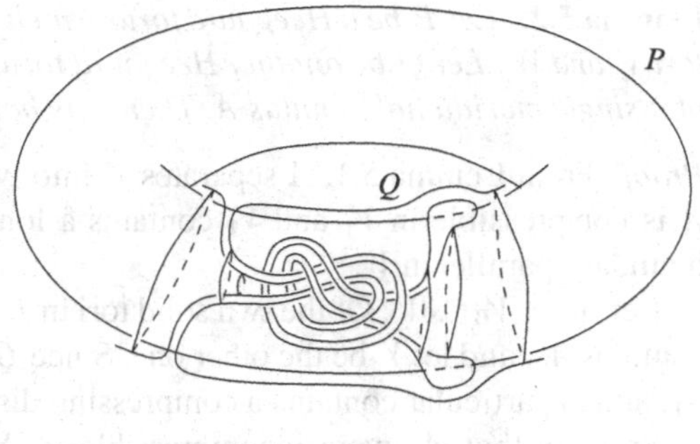

Fig. 5.2 Heegaard tori in very good position with non-boundary-parallel meridional annuli

circle of V. The boundary of a regular neighborhood of such a circle is a Heegaard torus of L. But here is one restriction on Heegaard tori:

Proposition 5.4. *Let P be a Heegaard torus in a lens space L, bounding solid tori V and W. If a loop ℓ embedded in P is a core circle for a solid torus of some genus-1 Heegaard splitting of L, then ℓ is a longitude for either V or W.*

Proof. Since L is not simply-connected, ℓ is not a meridian for either V or W, consequently $\pi_1(\ell) \to \pi_1(V)$ and $\pi_1(\ell) \to \pi_1(W)$ are injective. So $P - \ell$ is an open annulus separating $L - \ell$, making $\pi_1(L - \ell)$ a free product with amalgamation $\mathbb{Z} *_{\mathbb{Z}} \mathbb{Z}$. Since ℓ is a core circle, $\pi_1(L - \ell)$ is infinite cyclic, so at least one of the inclusions of the amalgamating subgroup to the infinite cyclic factors is surjective. □

Let F_1 and F_2 be transversely intersecting embedded surfaces in the interior of a three-manifold M. A component of $F_1 \cap F_2$ is called *discal* when it is contractible in both F_1 and F_2, and *biessential* when it is essential in both. We say that F_1 and F_2 are *in good position* when every component of their intersection is either discal or biessential, and at least one is biessential, and we say that they are *in very good position* when they are in good position and every component of their intersection is biessential.

Later, we will go to considerable effort to obtain pairs of Heegaard tori for lens spaces that intersect in very good position. Even then, the configuration can be complicated. Consider a Heegaard torus P bounding solid tori V and W, and another Heegaard torus Q that meets P in very good position. When the intersection circles are not meridians for either V or W, the components of $Q \cap V$ and $Q \cap W$ are annuli that are incompressible in V and W, and must be as described in Proposition 5.3. But if the intersection circles are meridians for one of the solid tori, say V, then $Q \cap V$ consists of meridional annuli, and as shown in Fig. 5.2, they need not be boundary-parallel. To obtain that configuration, one starts with a torus Q parallel to P and outside P, and changes Q by an isotopy that moves a meridian c of Q in a regular neighborhood of a meridian disk of P. First, c passes across a meridian in P, then shrinks down to a small circle which traces around a knot. Then, it expands out to another meridian in P and pushes across. The resulting torus meets P in four circles which are meridians for V, and meets V in two annuli, both isotopic to the non-boundary-parallel annulus in Fig. 5.1. The next lemma gives a small but important restriction on meridional annuli of $Q \cap V$.

Lemma 5.2. *Let P be a Heegaard torus which separates a lens space into two solid tori V and W. Let Q be another Heegaard torus whose intersection with V consists of a single meridional annulus A. Then A is boundary-parallel in V.*

Proof. From Lemma 5.1, A separates V into two components V_0 and V_1, such that A is compressible in V_1 and V_1 contains a longitude of V. Suppose that A is not boundary-parallel in V.

Let $A_0 = V_0 \cap \partial V$. Of the two solid tori in L bounded by Q, let X be the one that contains A_0, and let Y be the other one. Since $Q \cap V$ consists only of A, Y contains V_1, and in particular contains a compressing disk for A in V_1 and a longitude for V.

Suppose that A_0 were incompressible in X. Since A_0 is not parallel to A, it would be parallel to $\overline{\partial X - A}$. So V_0 would contain a core circle of X. Since $\pi_1(V_0) \to \pi_1(V)$ is the zero homomorphism, this implies that L is simply-connected, a contradiction. So A_0 is compressible in X. A compressing disk for A_0 in X is part of a 2-sphere that meets Y only in a compressing disk of A in V_1. This 2-sphere has algebraic intersection ± 1 with the longitude of V in V_1, contradicting the irreducibility of L. □

Regarding D^2 as the unit disk in the plane, for $0 < r < 1$ let rD^2 denote $\{(x, y) \mid x^2 + y^2 \leq r^2\}$. A solid torus X embedded in a solid torus V is called *concentric in* V if there is some product structure $V = D^2 \times S^1$ such that $X = rD^2 \times S^1$. Equivalently, X is in the interior of V and some (hence every) core circle of X is a core circle of V.

The next lemma shows how we will use Heegaard tori that meet in very good position.

Lemma 5.3. *Let P be a Heegaard torus which separates a lens space into two solid tori V and W. Let Q be another Heegaard torus, that meets P in very good position, and assume that the annuli of $Q \cap V$ are incompressible in V. Then at least one component C of $V - (Q \cap V)$ satisfies both of the following:*

1. *C is a core region for V.*
2. *Suppose that Q is moved by isotopy to a torus Q_1 in W, by pushing the annuli of $Q \cap V$ one-by-one out of V using isotopies that move them across regions of $V - C$, and let X be the solid torus bounded by Q_1 that contains V. Then V is concentric in X.*
3. *After all but one of the annuli have been pushed out of V, the image Q_0 of Q is isotopic to P relative to $Q_0 \cap P$.*

Proof. Assume first that $Q \cap V$ has only one component A. Then ∂A separates P into two annuli, A_1 and A_2. Since A is incompressible in V, it is parallel in V to at least one of the A_i, say A_1. Let $A' = Q \cap W$.

If A' is longitudinal, then A' is parallel in W to A_2. So pushing A across A_1 moves Q to a torus in W parallel to P, and the lemma holds, with C being the region between A and A_2. An isotopy from Q to P can be carried out relative to $Q \cap P$, giving the last statement of the lemma. Suppose that A' is not longitudinal. If A' is incompressible, then it is boundary parallel in W. If A' is not incompressible, then since P and Q meet in very good position, A' is meridional, and by Lemma 5.2 it is again boundary-parallel in W. If A' is parallel to A_2, then we are finished as before. If A' is parallel to A_1, but not to A_2, then there is an isotopy moving Q to a regular neighborhood of a core circle of A_1. By Proposition 5.4, A is longitudinal, so must also be parallel in V to A_2. In this case, we take C to be the region between A and A_1.

Suppose now that $Q \cap V$ and hence also $Q \cap W$ consist of n annuli, where $n > 1$. By isotopies pushing outermost annuli in V across P, we obtain Q_0 with $Q_0 \cap V$ consisting of one annulus A. At least one of its complementary components, call it C, satisfies the lemma. Let Z be the union of the regions across which the $n-1$ annuli were pushed. Since C is a core region, $C \cap (V - Z)$ is also a core region (since a core circle of V in C can be moved, by the reverse of the pushout isotopies, to a core circle of V in $C \cap (V - Z)$). So $C \cap (V - Z)$ satisfies the conclusion of the lemma. □

Here is a first consequence of Lemma 5.3.

Corollary 5.1. *Let P be a Heegaard torus which separates a lens space into two solid tori V and W, and let Q be another Heegaard torus separating it into X and Y. Assume that Q meets P in very good position. If the circles of $P \cap Q$ are meridians in X or in Y (respectively, in X and in Y), then they are meridians in V or in W (respectively, in V and in W). An analogous assertion holds for longitudes.*

Proof. We may choose notation so that the annuli of $Q \cap V$ are incompressible in V. Use Lemma 5.3 to move Q out of V. After all but one annulus has been pushed out, the image Q_0 of Q is isotopic to P relative to $Q_0 \cap P$. That is, the original Q is isotopic to P by an isotopy relative to $Q_0 \cap P$. If the circles of $Q \cap P$ were originally meridians of X or Y, then in particular those of $Q_0 \cap P$ are meridians of X or Y after the isotopy, that is, of V or W. The "and" assertion and the case of longitudes are similar. □

5.5 Sweepouts, and Levels in Very Good Position

In this section we will define sweepouts and related structures. Also, we will prove an important technical lemma concerning pairs of sweepouts having levels that meet in very good position.

By a *sweepout* of a closed orientable three-manifold, we mean a smooth map $\tau\colon P \times [0,1] \to M$, where P is a closed orientable surface, such that

1. $T_0 = \tau(P \times \{0\})$ and $T_1 = \tau(P \times \{1\})$ are disjoint graphs with each vertex of valence 3.
2. Each T_i is a union of a collection of smoothly embedded arcs and circles in M.
3. $\tau|_{P\times(0,1)}\colon P \times (0,1) \to M$ is a diffeomorphism onto $M - (T_0 \cup T_1)$.
4. Near $P \times \partial I$, τ gives a mapping cylinder neighborhood of $T_0 \cup T_1$.

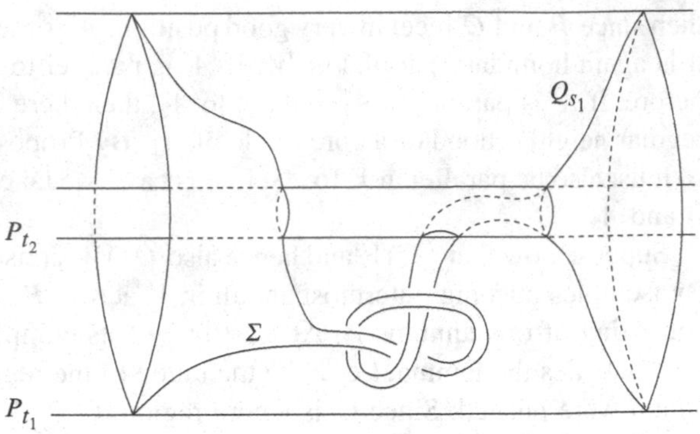

Fig. 5.3 Case (2) of Lemma 5.4

Associated to any t with $0 < t < 1$, there is a Heegaard splitting $M = V_t \cup W_t$, where $V_t = \tau(P \times [0,t])$ and $W_t = \tau(P \times [t,1])$. For each t, T_0 is a deformation retract of V_t and T_1 is a deformation retract of W_t. We denote $\tau(P \times \{t\})$ by P_t, and call it a *level* of τ. Also, for $0 < s < t < 1$ we denote $\tau(P \times [s,t])$ by $R(s,t)$. Note that any genus-1 Heegaard splitting of L provides sweepouts with T_0 and T_1 as core circles of the two solid tori, and the Heegaard torus as one of the levels.

A sweepout $\tau: P \times [0,1] \to M$ induces a continuous projection function $\pi: M \to [0,1]$ by the rule $\pi(\tau(x,t)) = t$. By composing this with a smooth bijection from $[0,1]$ to $[0,1]$ all of whose derivatives vanish at 0 and at 1, we may reparameterize τ to ensure that π is a smooth map. We always assume that τ has been selected to have this property.

By a *spine* for a closed connected surface P, we mean a one-dimensional cell complex in P whose complement consists of open disks.

The next lemma gives very strong restrictions on levels of two different sweepouts of a lens space that intersect in very good position.

Lemma 5.4. *Let L be a lens space. Let $\tau: T \times [0,1] \to L$ be a sweepout as above, where T is a torus. Let $\sigma: T \times [0,1] \to L$ be another sweepout, with levels $Q_s = \sigma(T \times \{s\})$. Suppose that for $t_1 < t_2$, $s_1 \neq s_2$, and $i = 1,2$, Q_{s_i} and P_{t_i} intersect in very good position, and that Q_{s_1} has no discal intersections with P_{t_2}. If Q_{s_1} has nonempty intersection with P_{t_2}, then either*

1. *Every intersection circle of Q_{s_1} with P_{t_2} is biessential, and consequently $Q_{s_1} \cap R(t_1,t_2)$ contains an annulus with one boundary circle essential in P_{t_1} and the other essential in P_{t_2}, or*
2. *For $i = 1,2$, $Q_{s_i} \cap P_{t_i}$ consists of meridians of W_{t_i}, and $Q_{s_1} \cap R(t_1,t_2)$ contains a surface Σ which is a homology from a circle of $Q_{s_1} \cap P_{t_1}$ to a union of circles in P_{t_2}.*

Figure 5.3 illustrates case (2) of Lemma 5.4.

We mention that to apply Lemma 5.4 when $t_1 > t_2$, we interchange the roles of V_{t_i} and W_{t_i}. The intersection circles in case (2) are then meridians of the V_{t_i} rather than the W_{t_i}.

Proof (of Lemma 5.4). Assume for now that the circles of $Q_{s_2} \cap P_{t_2}$ are not meridians of W_{t_2}.

We first rule out the possibility that there exists a circle of $Q_{s_1} \cap P_{t_2}$ that is inessential in Q_{s_1}. If so, there would be a circle C of $Q_{s_1} \cap P_{t_2}$, bounding a disk D in Q_{s_1} with interior disjoint from P_{t_2}. Since Q_{s_1} and P_{t_2} have no discal intersections, C is essential in P_{t_2}, so D is a meridian disk for V_{t_2} or W_{t_2}. It cannot be a meridian disk for V_{t_2}, for then some circle of $D \cap P_{t_1}$ would be a meridian of V_{t_1}, contradicting the fact that Q_{s_1} and P_{t_1} meet in very good position. But D cannot be a meridian disk for W_{t_2}, since D is disjoint from Q_{s_2} and the circles of $Q_{s_2} \cap P_{t_2}$ are not meridians of W_{t_2}.

We now rule out the possibility that there exists a circle of $Q_{s_1} \cap P_{t_2}$ that is essential in Q_{s_1} and inessential in P_{t_2}. There is at least one biessential intersection circle of Q_{s_1} with P_{t_1}, hence also an annulus A in Q_{s_1} with one boundary circle inessential in P_{t_2} and the other essential in either P_{t_1} or P_{t_2}, with no intersection circle of the interior of A with $P_{t_1} \cup P_{t_2}$ essential in A. The interior of A must be disjoint from P_{t_1}, since Q_{s_1} meets P_{t_1} in very good position. It must also be disjoint from P_{t_2}, by the previous paragraph. So, since A has at least one boundary circle in P_{t_2}, it is properly embedded either in $R(t_1,t_2)$ or in W_{t_2}. It cannot be in $R(t_1,t_2)$, since it has one boundary circle inessential in P_{t_2} and the other essential in $P_{t_1} \cup P_{t_2}$. So A is in W_{t_2}, and since one boundary circle is inessential in P_{t_2}, the other must be a meridian, contradicting the assumption that no circle of $Q_{s_2} \cap P_{t_2}$ is a meridian of W_{t_2}. Thus conclusion (1) holds when circles of $Q_{s_2} \cap P_{t_2}$ are not meridians of W_{t_2}.

Assume now that the circles of $Q_{s_2} \cap P_{t_2}$ are meridians of W_{t_2}. We will achieve conclusion (2).

Suppose first that some circle of $Q_{s_1} \cap P_{t_2}$ is essential in Q_{s_1}. Then there is an annulus A in Q_{s_1} with one boundary circle essential in P_{t_1}, the other essential in P_{t_2}, and all intersections of the interior of A with $P_{t_1} \cup P_{t_2}$ inessential in A. Since Q_{s_1} meets P_{t_1} in very good position, the interior of A must be disjoint from P_{t_1}. So $A \cap R(t_1,t_2)$ contains a planar surface Σ with one boundary component a circle of $Q_{s_1} \cap P_{t_1}$ and the other boundary components circles in P_{t_2} which are meridians in W_{t_2}, giving the conclusion (2) of the lemma.

Suppose now that every circle of $Q_{s_1} \cap P_{t_2}$ is contractible in Q_{s_1}. We will show that this case is impossible. An intersection circle innermost on Q_{s_1} bounds a disk D in Q_{s_1} which is a meridian disk for W_{t_2}, since ∂D is essential in P_{t_2} and disjoint from $Q_{s_2} \cap P_{t_2}$. Now, use Lemma 5.3 to push $Q_{s_2} \cap V_{t_2}$ out of V_{t_2} by an ambient isotopy of L. Suppose for contradiction that one of these pushouts, say, pushing an annulus A_0 in Q_{s_2} across an annulus in P_{t_2}, also eliminates a circle of $Q_{s_1} \cap P_{t_1}$. Let Z be the region of parallelism across which A_0 is pushed. Since Z contains an essential loop of Q_{s_1}, and each circle of $Q_{s_1} \cap P_{t_2}$ is contractible in Q_{s_1}, Z contains a spine of Q_{s_1}. This spine is isotopic in Z into a neighborhood of a boundary circle of A_0. Since this boundary circle is a meridian of W_{t_2}, every circle in the spine is contractible in L. This contradicts the fact that Q_{s_1} is a Heegaard torus. So the pushouts do not eliminate intersections of Q_{s_1} with P_{t_1}, and after the pushouts are completed, the image of Q_{s_1} still meets P_{t_1}.

During the pushouts, some of the intersection circles of Q_{s_1} with P_{t_2} may disappear, but not all of them, since the pushouts only move points into W_{t_2}. So after the pushouts, there is a circle of $Q_{s_1} \cap P_{t_2}$ that bounds a innermost disk in Q_{s_1} (since all the original intersection circles of Q_{s_1} with P_{t_2} bound disks in Q_{s_1}, and the new intersection circles are a subset of the old ones). Since the boundary of this disk is a meridian of W_{t_2}, the disk it bounds in Q_{s_1} must be a meridian disk of W_{t_2}. The image of Q_{s_2} lies in W_{t_2} and misses this meridian disk, contradicting the fact that Q_{s_2} is a Heegaard torus. □

5.6 The Rubinstein–Scharlemann Graphic

The purpose of this section is to present a number of definitions, and to sketch the proof of Theorem 5.2 below, originally from [58]. It requires the hypothesis that two sweepouts meet in general position in a strong sense that we call Morse general position. In Sect. 5.9, this proof will be adapted to the weaker concept of general position developed in Sect. 5.8.

Consider a smooth function $f: (\mathbb{R}^2, 0) \to (\mathbb{R}, 0)$. A critical point of f is *stable* when it is locally equivalent under smooth change of coordinates of the domain and range to $f(x, y) = x^2 + y^2$ or $f(x, y) = x^2 - y^2$. The first type is called a *center,* and the second a *saddle.* An unstable critical point is called a *birth-death* point if it is locally $f(x, y) = x^2 + y^3$.

Let $\tau: P \times [0, 1] \to M$ be a sweepout as in Sect. 5.5. As in that section, we denote $\tau(P \times \{0, 1\})$ by T, $\tau(P \times \{t\})$ by P_t, $\tau(P \times [0, t])$ by V_t, and $\tau(P \times [t, 1])$ by W_t. For a second sweepout $\sigma: Q \times [0, 1] \to M$, we denote $\sigma(Q \times \{0, 1\})$ by S, $\sigma(Q \times \{s\})$ by Q_s, $\sigma(Q \times [0, s])$ by X_s, and $\sigma(Q \times [s, 1])$ by Y_s. We call Q_s a *σ-level* and P_t a *τ-level.*

A tangency of Q_s and P_t at a point w is said to be *of Morse type* at w if in some local xyz-coordinates with origin at w, P_t is the xy-plane and Q_s is the graph of a function which has a stable critical point or a birth-death point at the origin. Note that this condition is symmetric in Q_s and P_t. We may refer to a tangency as stable or unstable, and as a center, saddle, or birth-death point.

A tangency of S with a τ-level is said to be *stable* if there are local xyz-coordinates in which the τ-levels are the planes $\mathbb{R}^2 \times \{z\}$ and S is the graph of $z = x^2$ in the xz-plane. In particular, the tangency is isolated and cannot occur at a vertex of S. There is an analogous definition of stable tangency of T with a σ-level.

We will say that σ and τ are in *Morse general position* when the following hold:

1. S is disjoint from T.
2. All tangencies of S with τ-levels and of T with σ-levels are stable.
3. All tangencies of σ-levels with τ-levels are of Morse type, and only finitely many are birth-death points.
4. Each pair consisting of a σ-level and a τ-level has at most two tangencies.

5. There are only finitely many pairs consisting of a σ-level and a τ-level with two tangencies, and for each of these pairs both tangencies are stable.

Suppose that P is a Heegaard surface in M, bounding a handlebody V. We define a *precompression* or *precompressing disk* for P in V to be an embedded disk D in M such that

1. ∂D is an essential loop in P.
2. D meets P transversely at ∂D, and V contains a neighborhood of ∂D.
3. The interior of D is transverse to P, and its intersections with P are discal.

Provided that M is irreducible, a precompression for P in V is isotopic relative to a neighborhood of ∂D to a compressing disk for P in V. In particular, if the Heegaard splitting is strongly irreducible, then the boundaries of a precompression for P in V and a precompression for P in $\overline{M - V}$ must intersect.

The following concept due to Casson and Gordon [9] is a crucial ingredient in [58]. A Heegaard splitting $M = V \cup_P W$ is called *strongly irreducible* when every compressing disk for V meets every compressing disk for W. A sweepout is called *strongly irreducible* when the associated Heegaard splittings are strongly irreducible. We can now state the main technical result of [58].

Theorem 5.2 (Rubinstein–Scharlemann). *Let $M \neq S^3$ be a closed orientable three-manifold, and let $\sigma, \tau \colon F \times [0,1] \to M$ be strongly irreducible sweepouts of M which are in Morse general position. Then there exists $(s,t) \in (0,1) \times (0,1)$ such that Q_s and P_t meet in good position.*

We will now review the proof of Theorem 5.2. The closure in I^2 of the set (s,t) for which Q_s and P_t have a tangency is a graph Γ. On ∂I^2, it can have valence-1 vertices corresponding to valence-3 vertices of S or T, and valence-2 vertices corresponding to points of tangency of S with a τ-level or T with a σ-level (see p. 1008 of [58], see also [41] for an exposition with examples). In the interior of I^2, it can have valence-4 vertices which correspond to a pair of levels which have two stable tangencies, and valence-2 vertices which correspond to pairs of levels having a birth-death tangency.

The components of the complement of Γ in the interior of I^2 are called *regions*. Each region is either unlabeled or bears a label consisting of up to four letters. The labels are determined by the following conditions on Q_s and P_t, which by transversality hold either for every (s,t) or for no (s,t) in a region.

1. If Q_s contains a precompression for P_t in V_t (respectively, in W_t), the region receives the letter A (respectively, B).
2. If P_t contains a precompression for Q_s in X_s (respectively, in Y_s), the region receives the letter X (respectively, Y).
3. If the region has neither an A-label nor a B-label, and V_t (respectively, W_t), contains a spine of Q_s, the region receives the letter b (respectively, a).
4. If the region has neither an X-label nor a Y-label, and X_s (respectively, Y_s), contains a spine of P_t, the region receives the letter y (respectively, x).

With these conventions, Q_s and P_t are in good position if and only if the region containing (s,t) is unlabeled. To check this, assume first that they are in good position. Since all intersections are biessential or discal, neither surface can contain a precompressing disk for the other, and since there is a biessential intersection circle, the complement of one surface cannot contain a spine for the other. For the converse, an intersection circle which is not biessential or discal leads to a precompression as in (1) or (2), so assume that all intersections are discal. Then the complement of the intersection circles in Q_s contains a spine, so the region has either an a- or b-label, and by the same reasoning applied to P_t the region has either an x- or y-label. This verifies the assertion, as well as the following lemma.

Lemma 5.5. *If the label of a region contains the letter a or b, then it must also contain either x or y. Similarly, if it contains x or y, then it must also contain a or b.*

We call the data consisting of the graph $\Gamma \subset I^2$ and the labeling of a subset of its regions the *Rubinstein–Scharlemann graphic* associated to the sweepouts. Regions of the graphic are called *adjacent* if there is an edge of Γ which is contained in both of their closures.

At this point, we begin to make use of the fact that the sweepouts are strongly irreducible. The labels will then have the following properties, where A stands for either of A and a, and B, X, and Y are defined similarly.

(RS1) A label cannot contain both an A and a B, or both an X and a Y (direct from the labeling rules and the definition of strong irreducibility).

(RS2) If the label of a region contains A, then the label of any adjacent region cannot contain B. Similarly for X and Y (Corollary 5.5 of [58]).

(RS3) If all four letters A, B, X, and Y appear in the labels of the regions that meet at a valence-4 vertex of Γ, then two opposite regions must be unlabeled (Lemma 5.7 of [58]).

Property (RS2) warrants special comment, since it will play a major role in our later work. The analysis of labels of adjacent regions given in Sect. 5 of [58] uses only the fact that for the points (s,t) in an open edge of Γ, the corresponding Q_s and P_t have a single stable tangency. The open edges of the more general graphics we will use for the diffeomorphisms in parameterized families in general position will still have this property, so the labels of their graphics will still satisfy property (RS2). They will not satisfy property (RS3), indeed the Γ for their graphics can have vertices of high valence, so property (RS3) will not even be meaningful.

We now analyze the labels of regions whose closures meet ∂I^2, as on p. 1012 of [58]. Consider first a region whose closure meets the side $s = 0$ (we consider s to be the horizontal coordinate, so this is the left-hand side of the square). The region must contains points (s,t) with s arbitrarily close to 0. These correspond to Q_s which are extremely close to S_0. For almost all t, S_0 is transverse to P_t, and for sufficiently small s any intersection of such a P_t with Q_s must be an essential circle of Q_s bounding a disk in P_t that lies in X_s, in which case the region must have an

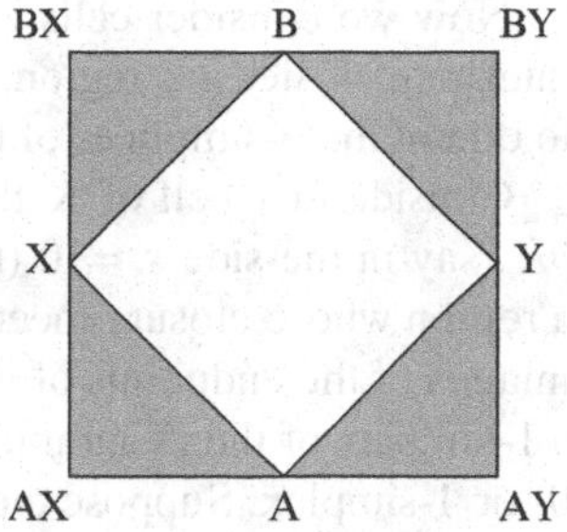

Fig. 5.4 The diagram

X-label. If P_t is disjoint from Q_s, then P_t lies in Y_s so the region has an x-label. That is, all such regions have an X-label. Similarly, the label of any region whose closure meets the edge $t = 0$ (respectively, $s = 1, t = 1$) contains A (respectively, Y, B).

We will set up some of the remaining steps a bit differently from those of [58], so that their adaptation to our later arguments will be more transparent. We have seen that it is sufficient to prove that there exists an unlabeled region in the graphic defined by the sweepouts. To accomplish this, Rubinstein and Scharlemann use the shaded subset of the square shown in Fig. 5.4. It is a simplicial complex in which each of the four triangles is a 2-simplex. Henceforth we will refer to it as *the Diagram.*

Suppose for contradiction that every region in the Rubinstein–Scharlemann graphic is labeled. Let Δ be a triangulation of I^2 such that each vertex of Γ and each corner of I^2 is a 0-simplex, and each edge of Γ is a union of 1-simplices. Let K be I^2 with the structure of a regular 2-complex dual to Δ. We observe the following properties of K:

(K1) Each 0-cell of K lies in the interior of a side of ∂I^2 or in a region.
(K2) Each 1-cell of K either lies in ∂I^2, or is disjoint from Γ, or crosses one edge of Γ transversely in one point.
(K3) Each 2-cell of K either contains no vertex of Γ, in which case all of its 0-cell faces that are not in ∂I^2 lie in one region or in two adjacent regions, or contains one vertex of Γ, in which case all of its 0-cell faces which do not lie in ∂I^2 lie in the union of the regions whose closures contain that vertex.

We now construct a map from K to the Diagram. First, each 0-cell in ∂K is sent to one of the single-letter 0-simplices of the diagram: if it lies in the side $s = 0$ (respectively, $t = 0$, $s = 1, t = 1$) then it is sent to the 0-simplex labeled X (respectively, A, Y, B). Similarly, any 1-cell in a side of ∂K is sent to the 0-simplex that is the image of its endpoints, and the four 1-cells in ∂K dual to the original corners are sent to the 1-simplex whose endpoints are the images of the endpoints of the 1-cell. Notice that ∂K maps essentially onto the circle consisting of the four diagonal 1-simplices of the Diagram.

We will now show that if there is no unlabeled region, this map extends to K, a contradiction. Since an unlabeled region produces pairs Q_s and P_t that meet in good position, this will complete the proof sketch of Theorem 5.2.

Now we consider cells of K that do not lie entirely in ∂K. Each 0-cell in the interior of K lies in a region. By (RS1), the label of each 0-cell has a form associated to one of the 0-simplices of the Diagram, and we send the 0-cell to that 0-simplex.

Consider a 1-cell of K that does not lie in ∂K. Suppose it has one endpoint in ∂K, say in the side $s = 0$ (the other cases are similar). The other endpoint lies in a region whose closure meets the side $s = 0$, so its label contains X. Therefore the images of the endpoints of the 1-cell both contain X, so lie either in a 0-simplex or a 1-simplex of the Diagram. We extend the map to the 1-cell by sending it into that 0- or 1-simplex. Suppose the 1-cell lies in the interior of K. Its endpoints lie either in one region or in two adjacent regions. If the former, or the latter and the labels of the regions are equal, we send the 1-cell to the 0-simplex for that label. If the latter and the labels of the regions are different, then property (RS2) shows that the labels span a unique 1-simplex of the Diagram, in which case we send the 1-cell to that 1-simplex.

Assuming that the map has been extended to the 1-cells in this way, consider a 2-cell of K. Suppose first that it has a face that meets the side $s = 0$ (the other cases are similar). Then each of its 0-cell faces lies in one of the sides $s = 0$, $t = 0$, or $t = 1$, or in a region whose closure meets $s = 0$. In the latter case, we have seen that the label of the region must contain X, so it cannot contain Y, and in particular it cannot be a single letter Y. In no case does the 0-cell map to the vertex Y of the Diagram, so the image of the boundary of the 2-cell maps into the complement of that vertex in the Diagram. Since that complement is contractible, the map extends over the 2-cell.

Suppose now that the 2-cell lies entirely in the interior of K. If it is dual to a 0-simplex of Δ that lies in a region or in the interior of an edge of Γ, then all its 0-cell faces lie in a region or in two adjacent regions. In this case, all of its one-dimensional faces map into some 1-simplex of the Diagram, so the map on the faces extends to a map of the 2-cell into that 1-simplex. Suppose the 2-cell is dual to a vertex of Γ. Its faces lie in the union of regions whose closures contain the vertex. If the vertex has valence 2, then all 0-cell faces lie in two adjacent regions (actually, in this case, the regions must have the same label) and the map extends to the 2-cell as before. If the vertex has valence 4, then by (RS3), the labels of the four regions whose closures contain the vertex must all avoid at least one of the four letters. This implies that the boundary of the 2-cell of K maps into a contractible subset of the Diagram. So again the map can be extended over the 2-cell, giving us the desired contradiction.

We emphasize that the map from K to the Diagram carries each 1-cell of K to a 0-simplex or a 1-simplex of the Diagram, principally due to property (RS2).

5.7 Graphics Having No Unlabeled Region

One cannot hope to perturb a parameterized family of sweepouts to be in Morse general position. One must allow for the possibility of levels having tangencies of high order, and having more than two tangencies. We will see in Sect. 5.8 that all

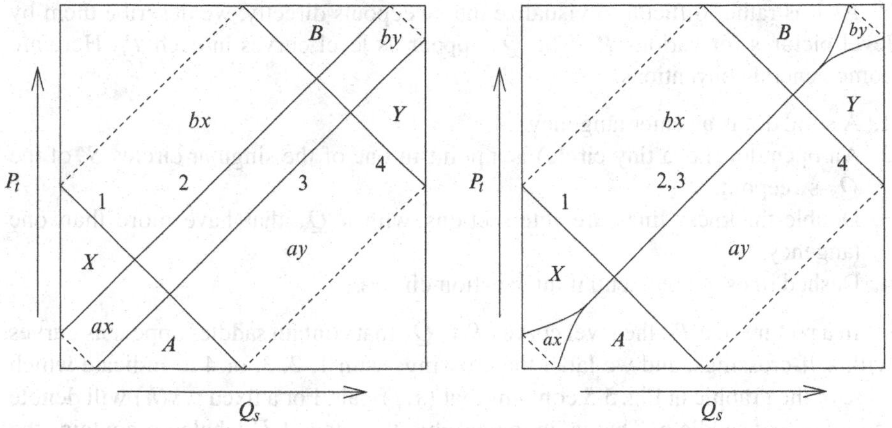

Fig. 5.5 Graphics before and after deformation

such phenomena can be isolated at the vertices of the graph Γ in the graphic. In particular, the (s,t) that lie on the open edges of Γ will still correspond to pairs of levels that have a single stable tangency, and therefore their associated graphics will still have property (RS2). Achieving this property for the edges of Γ will require considerable effort, so before beginning the task, we will show that the hard work really is necessary. We will give here an example of a pair of sweepouts on $S^2 \times S^1$ (that is, on $L(0,1)$) which have a graphic with no unlabeled region. It will be clear that what goes wrong is the existence of edges of Γ that consist of pairs having multiple tangencies, and the corresponding failure of the graphic to have property (RS2).

We do not have an explicit counterexample of this kind on a lens space, which would be even more complicated to describe, but we think it is fairly clear that the construction, which starts with a simple pair of sweepouts and "closes up" a good region in their graphic, could be carried out on a typical pair of sweepouts.

This section is not part of the proof of the Smale Conjecture for Lens Spaces, and can be read independently (provided that one is familiar with Rubinstein–Scharlemann graphics and their labeling scheme).

The first step is to construct a pair of sweepouts of $S^2 \times S^1$, with the graphic shown on the left in Fig. 5.5. In Fig. 5.5, the edges of pairs for which the corresponding levels have a single center tangency are shown as dotted. The four corner regions are not labeled, since their labels are the same as the regions that are adjacent to them along an edge of centers.

After constructing the sweepouts that produce the first graphic, we will see how to move one of the sweepouts by isotopy to "collapse" the unlabeled region. Two edges of the first graphic are moved to coincide, producing the graphic on the right in Fig. 5.5. The three open edges that lie on the diagonal $y = x$ consist of pairs of levels which have two saddle tangencies. The two vertices where the edges labeled 1 and 4 cross the diagonal at points corresponding to pairs having three saddle tangencies.

As it is rather difficult to visualize the sweepouts directly, we describe them by level pictures for various P_t. The Q_s appear as level curves in each P_t. Here are some general conventions:

1. A solid dot is a center tangency.
2. An open dot (i.e. a tiny circle) is a point in one of the singular circles S_i of the Q_s-sweepout.
3. Double-thickness lines are intersections with a Q_s that have more than one tangency.
4. Dashed lines are biessential intersection circles.

In a picture of a P_t, the level curves $P_t \cap Q_s$ that contain saddles appear as curves with self-crossings, and we label the crossings with 1, 2, 3, or 4 to indicate which edge of the graphic in Fig. 5.5 contains that (s,t)-pair. For a fixed t, $s(n)$ will denote the s-level of saddle n. That is, in the graphic the edge of Γ labeled n contains the point $(s(n),t)$.

Figure 5.6 shows some P_t with $t \leq 1/2$, for a sweepout of $S^2 \times S^1$ whose graphic is the one shown in the left of Fig. 5.5. Here are some notes on Fig. 5.6.

1. In (a)–(f), the circles $x =$ constant are longitudes of V_t, and the circles $y =$ constant are meridians.
2. The point represented by the four corners is the point of P_t with largest s-level. In (a) it is a tangency of $P_{1/2}$ with S_1, and in (b)–(f) it is a center tangency of P_t with $Q_{t+1/2}$.
3. The open dots in the interior of the squares are intersections of P_t with S_0. In (a) it is a tangency of $P_{1/2}$ with S_0, in (b)–(e) they are transverse intersections. In (f), P_t is disjoint from S_0.
4. In (b), saddle 1 has appeared. Circles of $Q_s \cap P_t$ with $s < s(1)$ are essential in Q_s, and these (s,t) lie in the region labeled X in the graphic. The level curve $P_t \cap Q_{s_1}$ that contains this saddle is the circle with one self-crossing in the middle of (b). Circles of $Q_s \cap P_t$ with $s_1/ < s < s_2/$ enclose this level curve. They are inessential in both Q_s and P_t, and these (s,t) lie in the region labeled bx. The vertical dotted lines are biessential intersections corresponding to a pair in the unlabeled region. Finally, one crosses $Q_{s(3)}$, and eventually reaches the center tangency.
5. The horizontal level curves shown in (f) are meridians of V_t that bound disks in the Q_s that contain them. This (s,t) lies in the region labeled A in the graphic.

For $t > 1/2$, the intersection pattern of P_t with the Q_s is isomorphic to the pattern for P_{1-t}, by an isomorphism for which Q_s corresponds to Q_{1-s}. As one starts t at $1/2$ and moves upward through t-levels, saddle 4 appears inside the component of $P_t - Q_{s(3)}$ that is an open disk, and expands until the level where $s(3) = s(4)$. The biessential intersection circles in (a)–(d) are again longitudes in V_t and in W_t, and the horizontal intersection circles in (f) are meridians of W_t. These (s,t) lie in the region labeled B in the graphic. This completes the description of the sweepouts in Morse general position.

Figure 5.7 shows some P_t for a sweepout of $S^2 \times S^1$ whose graphic is the one shown in the right of Fig. 5.5. This sweepout is obtained from the previous one by

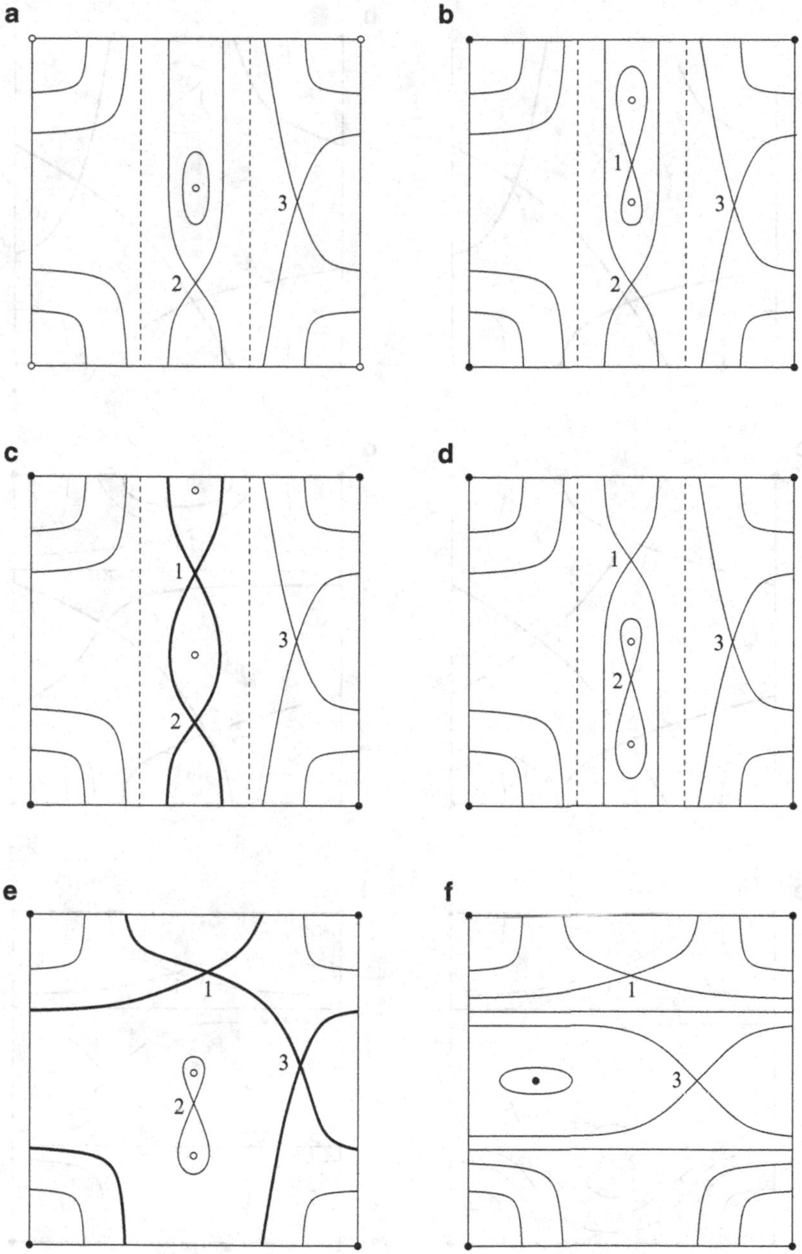

Fig. 5.6 Intersections of the Q_s with fixed P_t as t decreases from $1/2$ to 0, for the sweepouts with an unlabeled region. (**a**) $P_{1/2}$. (**b**) P_t where $s(1) < s(2) < s(3)$. (**c**) P_t where $s(1) = s(2)$. (**d**) P_t where $s(2) < s(1) < s(3)$. (**e**) P_t where $s(1) = s(3)$. (**f**) P_t where $s(3) < s(1)$, and after saddle 2 changes to a center

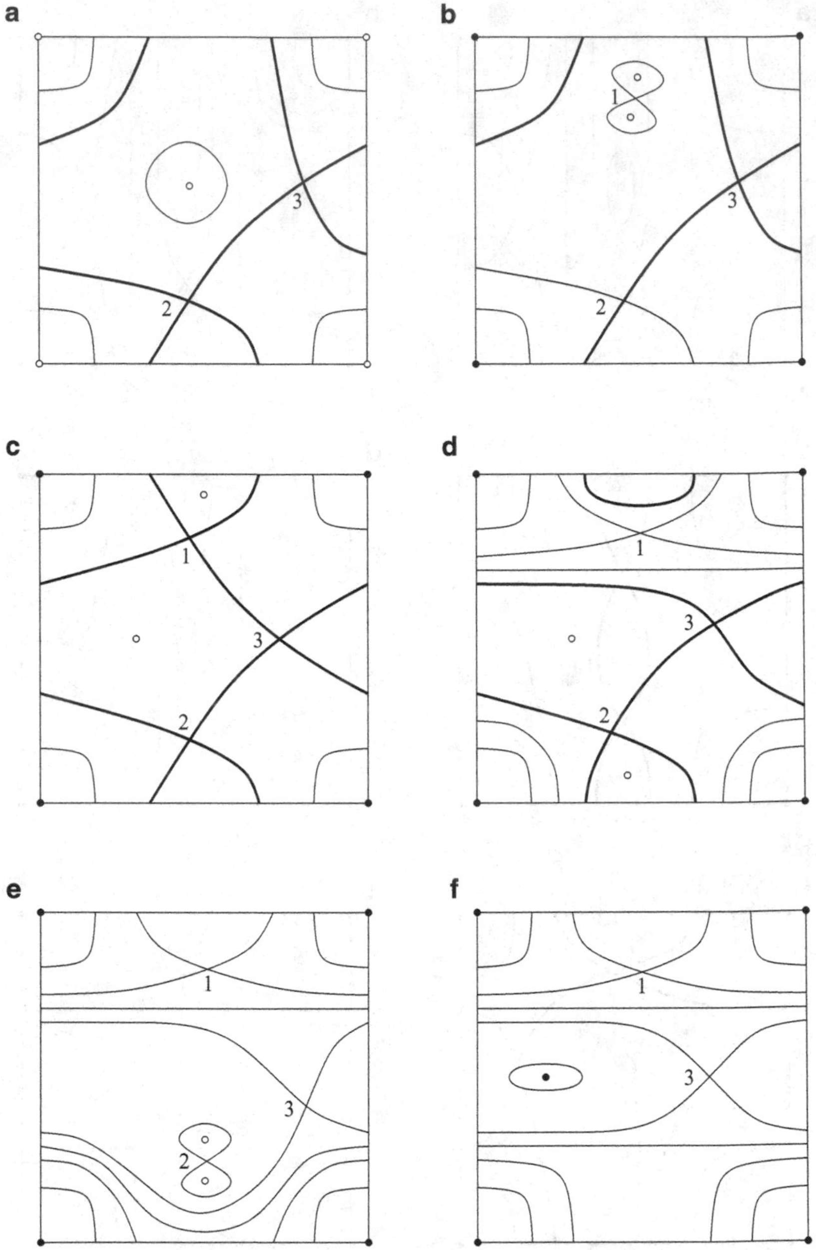

Fig. 5.7 Intersections of the Q_s with fixed P_t as t decreases from $1/2$ to 0, for the sweepouts with no unlabeled region. (**a**) $P_{1/2}$. (**b**) P_t where $s(1) < s(2) = s(3)$. (**c**) P_t where $s(1) = s(2) = s(3)$. (**d**) P_t where $s(2) = s(3) < s(1)$. (**e**) P_t where $s(2) < s(3) < s(1)$. (**f**) P_t where $s(3) < s(1)$, and after saddle 2 changes to a center

an isotopy that moves parts of the Q_s levels down (to lower t-levels) near saddle 2 and up near saddle 3. Again, the portion that is shown fits together with a similar portion for $1/2 \leq t \leq 1$. As t increases past $1/2$, saddle 4 appears in the component of $P_t - S_{s(2)}$ that contains the point which appears as the four corners.

5.8 Graphics for Parameterized Families

In this section we prove that a parameterized family of sweepouts can be perturbed so that a suitable graphic exists at each parameter. As discussed in Sect. 5.7, in a parameterized family one must allow for the possibility of levels having tangencies of high order, and having more than two tangencies.

Additional complications arise because one cannot avoid having parameters where the singular sets of the sweepouts intersect, or where the singular sets have high-order tangencies with levels. We sidestep these complications by working only with sweepout parameters that lie in an interval $[\epsilon, 1-\epsilon]$. The graphic is only considered to exist on the square $[\epsilon, 1-\epsilon] \times [\epsilon, 1-\epsilon]$, which we call I^2_ϵ. The number ϵ is chosen so that the labels of regions whose closure meets a side of I^2_ϵ will be known to include certain letters. Just as before, this will ensure that the map to the Diagram be essential on the boundary of the dual complex K.

These considerations motivate our definition of a general position family of diffeomorphisms. As usual, let M be a closed orientable three-manifold and $\tau\colon P \times [0,1] \to M$ a sweepout with singular set $T = T_0 \cup T_1$ and level surfaces P_t bounding handlebodies V_t and W_t. Let $f\colon M \times W \to M$ be a parameterized family of diffeomorphisms, where W is a compact manifold. For $u \in W$ we denote the restriction of f to $M \times \{u\}$ by f_u. When a choice of parameter u has been fixed, we denote $f_u(P_s)$ by Q_s, and $f_u(V_s)$ and $f_u(W_s)$ by X_s and Y_s respectively. When Q_s meets P_t transversely, a label is assigned to (s,t) as in Sect. 5.6.

A preliminary definition will be needed. We say that a positive number ϵ *gives border label control* for f if the following hold at each parameter u:

1. If $t \leq 2\epsilon$, then there exists r such that Q_r meets P_t transversely and contains a compressing disk of V_t.
2. If $t \geq 1-2\epsilon$, then there exists r such that Q_r meets P_t transversely and contains a compressing disk of W_t.
3. If $s \leq 2\epsilon$, then there exists r such that P_r meets Q_s transversely and contains a compressing disk of X_s.
4. If $s \geq 1-2\epsilon$, then there exists r such that P_r meets Q_s transversely and contains a compressing disk of Y_s.

Throughout this section, a *graph* is a compact space which is a disjoint union of a CW-complex of dimension ≤ 1 and circles. The circles, if any, are considered to be open edges of the graph.

We say that f is *in general position* (with respect to the sweepout τ) if there exists $\epsilon > 0$ such that ϵ gives border label control for f and such that the following hold for each parameter $u \in W$.

(GP1) For each (s,t) in I^2_ϵ, $Q_s \cap P_t$ is a graph. At each point in an open edge of this graph, Q_s meets P_t transversely. At each vertex, they are tangent.
(GP2) The $(s,t) \in I^2_\epsilon$ for which Q_s has a tangency with P_t form a graph Γ_u in I^2_ϵ.
(GP3) If (s,t) lies in an open edge of Γ_u, then Q_s and P_t have a single stable tangency.

The next lemma is immediate from the definition of border label control and the labeling rules for regions. It does not require that we be working with lens spaces, so we state it as a lemma with weaker hypotheses.

Lemma 5.6. *Suppose that $f: M \times W \to M$ is in general position with respect to τ. Assume that $M \neq S^3$ and that the Heegaard splittings associated to τ are strongly irreducible. Suppose that ϵ gives border label control for f.*

1. *If $t \leq \epsilon$, then the label of (s,t) contains A.*
2. *If $t \geq 1 - \epsilon$, then the label of (s,t) contains B.*
3. *If $s \leq \epsilon$, then the label of (s,t) contains X.*
4. *If $s \geq 1 - \epsilon$, then the label of (s,t) contains Y.*

Here is the main result of this section.

Theorem 5.3. *Let $f: M \times W \to M$ be a parameterized family of diffeomorphisms. Then by an arbitrarily small deformation, f can be put into general position with respect to τ.*

The proof of Theorem 5.3 will constitute the remainder of this section. Since the argument is rather long, we will break it into subsections. Until Sect. 5.8.6, M can be a closed manifold of arbitrary dimension m.

5.8.1 Weak Transversality

Although individual maps may be put transverse to a submanifold of the range, it is not possible to perturb a parameterized family so that each individual member of the family is transverse. But a very nice result Bruce, Theorem 1.1 of [8], allows one to simultaneously improve the members of a family.

Theorem 5.4 (J.W. Bruce). *Let A, B and U be smooth manifolds and $C \subset B$ a submanifold. There is a residual family of mappings $F \in C^\infty(A \times U, B)$ such that:*

(a) *For each $u \in U$, the restriction $F_u = F|_{A\times\{u\}}: A \to B$ is transverse to C except possibly on a discrete set of points.*

(b) *For each $u \in U$, the set $F_u^{-1}(C)$ is a smooth submanifold of codimension equal to the codimension of C in B, except possibly at a discrete set of points. At each of these exceptional points $F_u^{-1}(C)$ is locally diffeomorphic to the germ of an algebraic variety, with the exceptional point corresponding to an isolated singular point of the variety.*

That is, $F_u^{-1}(C)$ is smooth except at isolated points where it has topologically a nice cone-like structure. It is not assumed that any of the manifolds involved is compact.

Theorem 1.3 of [8] is a version of Theorem 5.4 in which C is replaced by a bundle $\phi: B \to D$. The statement is:

Theorem 5.5 (J.W. Bruce). *For a residual family of mappings $F \in C^\infty(A \times U, B)$, the conclusions of Theorem 5.4 hold for all submanifolds $C = \phi^{-1}(d)$, $d \in D$.*

We should comment on the significance of the residual subset in these two theorems. The method of proof of these theorems is to define, in an appropriate jet space, a locally algebraic subset which contains the jets of all the maps that fail these weak transversality conditions. These subsets have increasing codimension as higher-order jets are taken. A variant of Thom transversality (Lemma 1.6 of [8]) allows one to perturb a parameterized family of maps so that these jets are avoided and the conclusion holds. When A and W are compact, the image of $A \times W$ will lie in the open complement of the locally algebraic sets of sufficiently high codimension. Consequently, any map sufficiently close to the perturbed map will also satisfy the conclusions of the theorems. In all of our applications, the spaces involved will be compact, and *we tacitly assume that the result of any procedure holds on an open neighborhood of the perturbed map.*

We now adapt the methodology of Bruce to prove a version of Theorem 5.4 in which the submanifold C is replaced by the zero set of a nontrivial polynomial. We will prove it only for the case when $A = I$, although a more general version should be possible.

Proposition 5.5. *Let $P: \mathbb{R}^n \to \mathbb{R}$ be a nonzero polynomial and put $V = P^{-1}(0)$. Let W be compact. Then for all G in an open dense subset of $C^\infty(I \times W, \mathbb{R}^n)$, each $G_u^{-1}(V)$ is finite.*

Proof. Let $J_0^k(1,n)$ be the space of germs of degree-k polynomials from $(\mathbb{R}, 0)$ to $\mathbb{R}^n$; an element of $J_0^k(1,n)$ can be written as $(a_{1,0} + a_{1,1}t + \cdots + a_{1,k}t^k, \ldots, a_{n,0} + a_{n,1}t + \cdots + a_{n,k}t^k)$, so that $J_0^k(1,n)$ can be identified with $\mathbb{R}^{(k+1)n}$. Note that the jet space $J^k(I, \mathbb{R}^n)$ can be regarded as $I \times J_0^k(1,n)$, by identifying the jet of $\alpha: I \to \mathbb{R}^n$ at t_0 with the jet of $\alpha(t - t_0)$ at 0.

Define a polynomial map $P_*: J_0^k(1,n) \to J_0^k(1,1)$ by applying P to the n-tuple $(a_{1,0} + a_{1,1}t + \cdots + a_{1,k}t^k, \ldots, a_{n,0} + a_{n,1}t + \cdots + a_{n,k}t^k)$, and then taking only the terms up to degree k. The inverse image $P_*^{-1}(0)$ is the set of k-jets α in $\mathbb{R}^n$ such that $P(\alpha(0)) = 0$ and the first k derivatives of $P \circ \alpha$ vanish at $t = 0$, that is, the set of germs of paths that lie in V up to k^{th}-order.

Lemma 5.7. *If P is nonconstant, then as $k \to \infty$, the codimension of $P_*^{-1}(0)$ also goes to ∞.*

Proof (of Lemma 5.7). It suffices to show that the rank of the Jacobian of P_* goes to ∞ as $k \to \infty$. For notational simplicity, we will give the proof for $P(X,Y)$, and it will be evident how the argument extends to the general case.

Write $a = a_0 + a_1 t + a_2 t^2 + \cdots$ and $b = b_0 + b_1 t + b_2 t^2 + \cdots$, and examine $P(a,b)$. We have $P_*(a,b) = Q_0 + Q_1 t + Q_2 t^2 + \cdots$ where each Q_i is a (finite) polynomial in $\mathbb{R}[a_0, b_0, a_1, b_1, \ldots]$. Notice that $Q_j = \frac{1}{j!}\frac{\partial^j P_*}{\partial t^j}\bigg|_{t=0}$.

It is instructive to calculate a few derivatives of $P_*(a,b)$. We have

$$\begin{aligned}
\frac{\partial P_*}{\partial t} &= a' P_X + b' P_Y \\
\frac{\partial^2 P_*}{\partial t^2} &= a'' P_X + b'' P_Y + (a')^2 P_{XX} + 2a'b' P_{XY} + (b')^2 P_{YY} \\
\frac{\partial^3 P_*}{\partial t^3} &= a''' P_X + b''' P_Y + a''a' P_{XX} + (a''b' + a'b'') P_{XY} + b''b' P_{YY} \\
&\quad + 2a'a'' P_{XX} + (2a''b' + 2a'b'') P_{XY} + 2b'b'' P_{YY} \\
&\quad + (a')^3 P_{XXX} + 3(a')^2 b' P_{XXY} + 3a'(b')^2 P_{XYY} + (b')^3 P_{YYY} \\
&= a''' P_X + b''' P_Y + 3a''a' P_{XX} + 3(a''b' + a'b'') P_{XY} + 3b''b' P_{YY} \\
&\quad + (a')^3 P_{XXX} + 3(a')^2 b' P_{XXY} + 3a'(b')^2 P_{XYY} + (b')^3 P_{YYY}
\end{aligned}$$

and at $t = 0$ these become

$$\begin{aligned}
Q_1 &= a_1 P_X(a_0,b_0) + b_1 P_Y(a_0,b_0) \\
2!Q_2 &= 2a_2 P_X(a_0,b_0) + 2b_2 P_Y(a_0,b_0) \\
&\quad + a_1^2 P_{XX}(a_0,b_0) + 2a_1 b_1 P_{XY}(a_0,b_0) + b_1^2 P_{YY}(a_0,b_0) \\
3!Q_3 &= 6a_3 P_X(a_0,b_0) + 6b_3 P_Y(a_0,b_0) + 6a_1 a_2 P_{XX}(a_0,b_0) \\
&\quad + 6(a_2 b_1 + a_1 b_2) P_{XY}(a_0,b_0) + 6b_1 b_2 P_{YY}(a_0,b_0) + a_1^3 P_{XXX}(a_0,b_0) \\
&\quad + 3a_1^2 b_1 P_{XXY}(a_0,b_0) + 3a_1 b_1^2 P_{XYY}(a_0,b_0) + b_1^3 P_{YYY}(a_0,b_0)
\end{aligned}$$

Induction shows that in general, writing $K_{rX,sX}$ for $\frac{\partial^{r+s} P}{\partial^r X \partial^s Y}(a_0,b_0)$, there are positive constants $c_{i_1\cdots i_r j_1 \cdots j_s}$ so that for large N,

$$Q_N = \sum K_{rX,sY}\Big(\sum c_{i_1\cdots i_r j_1\cdots j_s} a_{i_1}\cdots a_{i_r} b_{j_1}\cdots b_{j_s}\Big). \tag{5.1}$$

For large N, all partial derivatives $K_{rX,sY}$ of P at (a_0,b_0) appear, and some must be nonzero since P is a polynomial. Notice also that there is no cancellation due to values of the $K_{rX,sY}$, since each monomial term $a_{i_1}\cdots a_{i_r}b_{j_1}\cdots b_{j_s}$ appears just once.

Any given a_i appears in some of the monomial terms of Q_N for all sufficiently large N. On the other hand, Q_N contains no a_i or b_i with $i > N$, so $\frac{\partial Q_i}{\partial a_j}$ vanishes for $j > i$, and similarly for $\frac{\partial Q_j}{\partial b_i}$. Therefore if we truncate at t^k, the Jacobian $\left[\left(\frac{\partial A_i}{\partial a_j}\right)\left(\frac{\partial A_i}{\partial b_j}\right)\right]$ is a $(k+1)\times 2(k+1)$ matrix consisting of (two, since we are in the case of a two-variable $P(X,Y)$) upper triangular blocks:

$$\begin{bmatrix} * & 0 & 0 & \cdots & 0 & * & 0 & 0 & \cdots & 0 \\ * & * & 0 & \cdots & 0 & * & * & 0 & \cdots & 0 \\ & & & \ddots & & & & & \ddots & \\ * & * & & \cdots & 0 & * & * & & \cdots & 0 \\ * & * & & \cdots & * & * & * & & \cdots & * \end{bmatrix}.$$

If the lemma is false, then there is some maximal rank of these Jacobians as $k\to\infty$. That is, there are, say, m rows such that every row is an $\mathbb{R}$-linear combination of these rows. For values of k much larger than m, all of these m rows have zeros in the upper triangular part of the two blocks. On the other hand, Eq. (5.1) and the observations that follow it show that for each fixed j, $\frac{\partial A_i}{\partial a_j}$ is nonzero for sufficiently large i. This completes the proof of Lemma 5.7. □

For each k, put $Z_k = P_*^{-1}(0)$. Lemma 5.7 shows that the codimension of Z_k in $J_0^k(1,n)$ goes to ∞ as $k\to\infty$. If $\alpha\colon(\mathbb{R},0)\to\mathbb{R}^n$ is a germ of a smooth map, and 0 is a limit point of $\alpha^{-1}(V)$, then all derivatives of $P\circ\alpha$ vanish at 0. That is, the k-jet of α at $t=0$ is contained in Z_k for every k.

By Lemma 1.6 of [8], there is a residual set of maps $G\in C^\infty(I\times W,\mathbb{R}^n)$ such that the jet extensions $j^kG\colon I\times W\to J^k(I,\mathbb{R}^n)$ defined by $j^kG(t,u)=j^kG_u(t)$ are transverse to $I\times Z_k$. For $k+1$ larger than the dimension of $I\times W$, this says that no point of $G_u^{-1}(0)$ is a limit point, so each $G_u^{-1}(0)$ is finite. □

5.8.2 *Finite Singularity Type*

In preparation for our later work, we will review some ideas from singularity theory. Let $g\colon(\mathbb{R}^m,0)\to(\mathbb{R}^p,0)$ be a germ of a smooth map. There is a concept of *finite singularity type* for g, whose definition is readily available in the literature (for example, [8, p. 117]). The basic idea of the proof of Theorem 5.4 (given as Theorem 1.1 in [8]) is to regard the submanifold C locally as the inverse image of 0 under a submersion s, then to perturb f so that for each u, the critical points of $s\circ f_u$

are of finite singularity type. In fact, this is exactly the definition of what it means for f_u to be weakly transverse to C. In particular, when C is a point, the submersion can be taken to be the identity, so we have:

Proposition 5.6. *Let $f: M \to \mathbb{R}$ be smooth. If f is weakly transverse to a point $r \in \mathbb{R}$, then at each critical point in $f^{-1}(r)$, the germ of f has finite singularity type.*

Let f and g be germs of smooth maps from $(\mathbb{R}^m, a)$ to $(\mathbb{R}^p, f(a))$. They are said to be *$\mathscr{A}$-equivalent* if there exist a germ φ_1 of a diffeomorphism of $(\mathbb{R}^m, a)$ and a germ φ_2 of a diffeomorphism of $(\mathbb{R}^p, f(a))$ such that $g = \varphi_2 \circ f \circ \varphi_1$. If φ_2 can be taken to be the identity, then f and g are called *$\mathscr{R}$-equivalent* (for *right-equivalent*). There is also a notion of contact equivalence, denoted by $\mathscr{K}$-equivalence, whose definition is readily available, for example in [72]. It is implied by $\mathscr{A}$-equivalence.

We use $j^k f$ to denote the k-jet of f; for fixed coordinate systems at points a and $f(a)$ this is just the Taylor polynomial of f of degree k. For $\mathscr{G}$ one of $\mathscr{A}$, $\mathscr{K}$, or $\mathscr{R}$, one says that f is finitely $\mathscr{G}$-determined if there exists a k so that any germ g with $j^k g = j^k f$ must be $\mathscr{G}$-equivalent to f. In particular, if f is finitely $\mathscr{G}$-determined, then for any fixed choice of coordinates at a and $f(a)$, f is $\mathscr{G}$-equivalent to a polynomial.

The elaborate theory of singularities of maps from $\mathbb{R}^m$ to $\mathbb{R}^p$ simplifies considerably when $p = 1$.

Lemma 5.8. *Let f be the germ of a map from $(\mathbb{R}^m, 0)$ to $(\mathbb{R}, 0)$, with 0 is a critical point of f. The following are equivalent.*

(i) *f has finite singularity type.*
(ii) *f is finitely $\mathscr{A}$-determined.*
(iii) *f is finitely $\mathscr{R}$-determined.*
(iv) *f is finitely $\mathscr{K}$-determined.*

Proof. In all dimensions, f is finitely $\mathscr{K}$-determined if and only if it is of finite singularity type (Corollary III.6.9 of [16], or alternatively the definition of finite singularity type of Bruce [8, p. 117] is exactly the condition given in Proposition (3.6)(a) of Mather [45] for f to be finitely $\mathscr{K}$-determined). Therefore (i) is equivalent to (iv). Trivially (ii) implies (iii), and (iii) implies (iv), and by Corollary 2.13 of [72], (iv) implies (ii). □

5.8.3 Semialgebraic Sets

Recall (see for example Chap. I.2 of [16]) that the class of *semialgebraic* subsets of $\mathbb{R}^m$ is defined to be the smallest Boolean algebra of subsets of $\mathbb{R}^m$ that contains all sets of the form $\{x \in \mathbb{R}^m \mid p(x) > 0\}$ with p a polynomial on $\mathbb{R}^m$. The collection of semialgebraic subsets of $\mathbb{R}^m$ is closed under finite unions, finite intersections, products, and complementation. The inverse image of a semialgebraic set under a polynomial mapping is semialgebraic. A nontrivial fact is the Tarski–Seidenberg Theorem (Theorem II.2(2.1) of [16]), which says that a polynomial image of a semialgebraic set is a semialgebraic set. Here is an easy lemma that we will need later.

Lemma 5.9. *Let S be a semialgebraic subset of $\mathbb{R}^n$. If S has empty interior, then S is contained in the zero set of a nontrivial polynomial in $\mathbb{R}^n$.*

Proof. Since the union of the zero sets of two polynomials is the zero set of their product, it suffices to consider a semialgebraic set of the form $(\cap_{i=1}^{r}\{x \mid p_i(x) \geq 0\}) \cap (\cap_{j=1}^{s}\{x \mid q_j(x) > 0\})$ where p_i and q_j are nontrivial polynomials. We will show that if S is of this form and has empty interior, then $r \geq 1$ and S is contained in the zero set of $\prod_{i=1}^{r} p_i$. Suppose that $x \in S$ but all $p_i(x) > 0$. Since all $q_j(x) > 0$ as well, there is an open neighborhood of x on which all p_i and all q_j are positive. But then, S has nonempty interior. □

5.8.4 The Codimension of a Real-Valued Function

It is, of course, fundamentally important that the Morse functions form an open dense subset of $C^\infty(M,\mathbb{R})$. But a great deal can also be said about the non-Morse functions. There is a "natural" stratification of $C^\infty(M,\mathbb{R})$ by subsets $\mathscr{F}_i$, where stratification here means that the $\mathscr{F}_i$ are disjoint subsets such that for every n the union $\cup_{i=0}^{n}\mathscr{F}_i$ is open. The functions in $\mathscr{F}_n$ are those of "codimension" n, which we will define below. In particular, $\mathscr{F}_0$ is exactly the open dense subset of Morse functions.

The union $\cup_{i=0}^{\infty}\mathscr{F}_i$ is not all of $C^\infty(M,\mathbb{R})$. However, the residual set $C^\infty(M,\mathbb{R}) - \cup_{i=0}^{\infty}\mathscr{F}_i$ is of "infinite codimension," and any parameterized family of maps $F\colon M \times U \to \mathbb{R}$ can be perturbed so that each F_u is of finite codimension. In fact, by applying Theorem 5.5 to the trivial bundle $1_{\mathbb{R}}\colon \mathbb{R} \to \mathbb{R}$ and noting Proposition 5.6, we may perturb any parameterized family so that each F_u is of finite singularity type at each of its critical points. The definition of $f \in C^\infty(M,\mathbb{R})$ being of finite codimension, given below, is exactly equivalent to the algebraic condition given in (3.5) of Mather [45] for f to be finitely $\mathscr{A}$-determined at each of its critical points (as noted in [45], this part of (3.5) was first due to Tougeron [67,68]). By Lemma 5.8, this is equivalent to f having finite singularity type at each of its critical points. We summarize this as

Proposition 5.7. *A map $f \in C^\infty(M,\mathbb{R})$ is of finite codimension if and only if it has finite singularity type at each of its critical points.*

We now recall material from Sect. 7 of [63]. Denote the smooth sections of a bundle E over M by $\Gamma(E)$. Until we reach Theorem 5.7, we will denote $C^\infty(M,\mathbb{R})$ by $C(M)$. For a compact subset $K \subset \mathbb{R}$, define $\mathrm{Diff}_K(\mathbb{R})$ to be the diffeomorphisms of $\mathbb{R}$ supported on K.

Fix an element $f \in C(M)$ and a compact subset $K \subset \mathbb{R}$ for which $f(M)$ lies in the interior of K. Define $\Phi\colon \mathrm{Diff}(M) \times \mathrm{Diff}_K(\mathbb{R}) \to C(M)$ by $\Phi(\varphi_1, \varphi_2) = \varphi_2 \circ f \circ \varphi_1$. The differential of Φ at $(1_M, 1_{\mathbb{R}})$ is defined by $D(\xi_1, \xi_2) = f_*\xi_1 + \xi_2 \circ f$. Here, $\xi_1 \in \Gamma(TM)$, which is regarded as the tangent space at 1_M of $\mathrm{Diff}(M)$, $\xi_2 \in \Gamma_K(T\,\mathbb{R})$, similarly identified with the tangent space at $1_{\mathbb{R}}$ of $\mathrm{Diff}_K(\mathbb{R})$, and $f_*\xi_1 + \xi_2 \circ f$ is regarded as an element of $\Gamma(f^*T\,\mathbb{R})$, which is identified with $C(M)$. The *codimension* $\mathrm{cdim}(f)$ of f is defined to be the real codimension of the image of D in $C(M)$. As will be seen shortly, the codimension of f tells the real codimension of the $\mathrm{Diff}(M) \times \mathrm{Diff}_K(\mathbb{R})$-orbit of f in $C(M)$.

Suppose that f has finite codimension c. In Sect. 7.2 of [63], a method is given for computing $\mathrm{cdim}(f)$ using the critical points of f. Fix a critical point a of f, with critical value $f(a) = b$. Consider $D_a\colon \Gamma_a(TM) \times C_b(\mathbb{R}) \to C_a(M)$, where a subscript as in $\Gamma_a(TM)$ indicates the germs at a of $\Gamma(TM)$, and so on. Notice that the codimension of the image of D_a is finite, indeed it is at most c.

Let A denote the ideal $f_*\Gamma_a(TM)$ of $C_a(M)$. This can be identified with the ideal in $C_a(M)$ generated by the partial derivatives of f. An argument using Nakayama's Lemma [63, p. 645] shows that A has finite codimension in $C_a(M)$, and that some power of $f(x) - f(a)$ lies in A. Define $\mathrm{cdim}(f, a)$ to be the dimension of $C_a(M)/A$, and $\dim(f, a, b)$ to be the smallest k such that $(f(x) - f(a))^k \in A$.

Here is what these are measuring. The ideal A tells what local deformations of f at a can be achieved by precomposing f with a diffeomorphism of M (near 1_M), thus $\mathrm{cdim}(f, a)$ measures the codimension of the $\mathrm{Diff}(M)$-orbit of the germ of f at a. The additional local deformations of f at a that can be achieved by postcomposing with a diffeomorphism of $\mathbb{R}$ (again, near $1_{\mathbb{R}}$) reduce the codimension by k, basically because Taylor's theorem shows that the germ at a of any $\xi_2(f(x))$ can be written in terms of the powers $(f(x) - f(a))^i$, $i < k$, plus a remainder of the form $K(x)(f(x) - f(a))^k$, which is an element of the ideal A. Thus $\mathrm{cdim}(f, a) - \dim(f, a, b)$ is the codimension of the image of D_a. For a noncritical point or a stable critical point such as $f(x, y) = x^2 - y^2$ at $(0, 0)$, this local codimension is 0, but for unstable critical points it is positive.

Now, let $\dim(f, b)$ be the maximum of $\dim(f, a, b)$, taken over the critical points a such that $f(a) = b$ (put $\dim(f, b) = 0$ if b is not a critical value). The codimension of f is then $\sum_{a \in M} \mathrm{cdim}(f, a) - \sum_{b \in \mathbb{R}} \dim(f, b)$.

Here is what is happening at each of the finitely many critical values b of f. Let $a_1, \ldots, a_\ell$ be the critical points of f with $f(a_i) = b$, and for each i write f_i for the germ of $f - f(a_i)$ at a_i. Consider the element $(f_1, \ldots, f_\ell) \in C_{a_1}(M)/A_1 \oplus \cdots \oplus C_{a_\ell}(M)/A_\ell$. The integer $\dim(f, b)$ is the smallest power of $(f_1, \ldots, f_\ell)$ that is trivial in $C_{a_1}(M)/A_1 \oplus \cdots \oplus C_{a_\ell}(M)/A_\ell$. The sum $\sum_i \mathrm{cdim}(f, a_i)$ counts how much codimension of f is produced by the inability to achieve local deformations of f near the a_i by precomposing with local diffeomorphisms at the a_i. This codimension is reduced by $\dim(f, b)$, because the germs of the additional deformations that can be achieved by postcomposition with diffeomorphisms of $\mathbb{R}$ near b are the linear combinations of $(1, \ldots, 1)$, $(f_1, \ldots, f_\ell)$, $(f_1^2, \ldots, f_\ell^2), \ldots$, $(f_1^{k-1}, \ldots, f_\ell^{k-1})$. Thus the contribution to the codimension from the critical points that map to b is $\sum_i \text{cdim}(f, a_i) - \dim(f, b)$, and summing over all critical values gives the codimension of f.

5.8.5 The Stratification of $C^\infty(M, \mathbb{R})$ by Codimension

The functions whose codimension is finite and equal to n form the stratum $\mathscr{F}_n$. In particular, $\mathscr{F}_0$ are the Morse functions, $\mathscr{F}_1$ are the functions either having all critical points stable and exactly two with the same critical value, or having distinct critical values and all critical points stable except one which is a birth-death point. Moving to higher strata occurs either from more critical points sharing a critical value, or from the appearance of more singularities of positive but still finite local codimension.

We use the natural notations $\mathscr{F}_{\geq n}$ for $\cup_{i \geq n} \mathscr{F}_i$, $\mathscr{F}_{>n}$ for $\cup_{i>n} \mathscr{F}_i$, and so on. In particular, $\mathscr{F}_{\geq 0}$ is the set of all elements of $C(M)$ of finite codimension, and $\mathscr{F}_{>0}$ is the set of all elements of finite codimension that are not Morse functions.

The main results of [63] (in particular, Theorems 8.1.1 and 9.2.4) show that the Sergeraert stratification is locally trivial, in the following sense.

Theorem 5.6 (Sergeraert). *Suppose that $f \in \mathscr{F}_n$. Then there is a neighborhood V of f in $C(M)$ of the form $U \times \mathbb{R}^n$, where*

1. *U is a neighborhood of* 1 *in* Diff$(M) \times$ Diff$_K(\mathbb{R})$.
2. *There is a stratification $\mathbb{R}^n = \cup_{i=0}^n F_i$, such that $\mathscr{F}_i \cap V = U \times F_i$.*

The inner workings of this result are as follows. Select elements $f_1, \ldots, f_n \in C(M)$ that represent a basis for the quotient of $C(M)$ by the image of the differential D of Φ at $(1_M, 1_\mathbb{R})$. For $x = (x_1, \ldots, x_n) \in \mathbb{R}^n$, the function $g_x = f + \sum_{i=1}^n x_i f_i$ is an element of $C(M)$. If the x_i are chosen in a sufficiently small ball around 0, which is again identified with $\mathbb{R}^n$, then these g_x form a copy E of $\mathbb{R}^n$ "transverse" to the image of Φ. Then, F_i is defined to be the intersection $E \cap \mathscr{F}_i$. A number of subtle results on this local structure and its relation to the action of Diff$(M) \times$ Diff$_K(\mathbb{R})$ are obtained in [63], but we will only need the local structure we have described here.

We remark that F_n is not necessarily just $\{0\} \in \mathbb{R}^n$, that is, the orbit of f under Diff$(M) \times$ Diff$_K(\mathbb{R})$ might not fill up the stratum $\mathscr{F}_n$ near f. This result, due to Hendriks [29], has been interpreted as saying that the Sergeraert stratification of $C(M)$ is *not* locally trivial (a source of some confusion), or that it is "pathological" (which we find far too pejorative).

Denoting $\cup_{i \geq 1} F_i$ by $F_{\geq 1}$, we have the following key technical result.

Proposition 5.8. *For some coordinates on E as $\mathbb{R}^n$, there are a neighborhood L of 0 in $\mathbb{R}^n$ and a nonzero polynomial p on $\mathbb{R}^n$ such that $p(L \cap F_{\geq 1}) = 0$.*

Proof. We will begin with a rough outline of the proof. Using Lemma 5.8, we may choose local coordinates at the critical points of f for which f is polynomial near each critical point. We will select the f_i in the construction of the transverse slice $E = \mathbb{R}^n$ to be polynomial on these neighborhoods. Now $F_{\geq 1}$ consists exactly of the choices of parameters x_i for which $f + \sum x_i f_i$ is not a Morse function, since they are the intersection of E with $\mathscr{F}_{\geq 1}$. We will show that they form a semialgebraic set. But $F_{\geq 1}$ has no interior, since otherwise (using Theorem 5.6) the subset of Morse functions $\mathscr{F}_0$ would not be dense in $C(M)$. So Lemma 5.9 shows that $F_{\geq 1}$ lies in the zero set of some nontrivial polynomial.

Now for the details. Recall that m denotes the dimension of M. Consider a single critical value b, and let $a_1, \dots, a_\ell$ be the critical points with $f(a_i) = b$. Fix coordinate neighborhoods U_i of the a_i with disjoint closures, so that a_i is the origin 0 in U_i. By Lemma 5.8, f is finitely $\mathscr{R}$-determined near each critical point, so on each U_i there is a germ φ_i of a diffeomorphism at 0 so that $f \circ \varphi_i$ is the germ of a polynomial. That is, by reducing the size of the U_i and changing the local coordinates, we may assume that on each U_i, f is a polynomial p_i. As explained in Sect. 5.8.4, the contribution to the codimension of f from the a_i is the dimension of the quotient

$$Q_b = \left(\oplus_{i=1}^{\ell} C_{a_i}(U_i)/A_i \right)/B$$

where B is the vector subspace spanned by

$$\{1, (p_1(x) - b, \dots, p_\ell(x) - b), \dots, ((p_1(x) - b)^{k-1}, \dots, (p_\ell(x) - b)^{k-1})\}\,.$$

Choose $q_{i,j}$, $1 \leq j \leq n_i$, where $q_{i,j}$ is a polynomial on U_i, so that the germs of the $q_{i,j}$ form a basis for Q_b. Fix vector spaces $\Lambda_i \cong \mathbb{R}^{n_i} = \{(x_{i,1}, \dots, x_{i,n_i})\}$; these will eventually be some of the coordinates on E.

In each U_i, select round open balls V_i and W_i centered at 0 so that $W_i \subset \overline{W_i} \subset V_i \subset \overline{V_i} \subset U_i$. We select them small enough so that the closures in $\mathbb{R}$ of their images under f do not contain any critical value except for b. Fix a smooth function $\mu\colon M \to [0, 1]$ which is 1 on $\cup \overline{W_i}$ and is 0 on $M - \cup V_i$, and put $f_{i,j} = \mu \cdot q_{i,j}$, a smooth function on all of M. Now choose a product $L = \prod_i L_i$, where each L_i is a round open ball centered at 0 in Λ_i, small enough so that if each $(x_{i,1}, \dots, x_{i,n_i}) \in L_i$, then each critical point of $f + \sum x_{i,j} f_{i,j}$ either lies in $\cup W_i$, or is one of the original critical points of f lying outside of $\cup U_i$.

We repeat this process for each of the finitely many critical values of f, choosing additional W_i and L_i so small that all critical points of $f + \sum x_{i,j} f_{i,j}$ lie in $\cup W_i$. That is, these perturbations of f are so small that each of the original critical points of f breaks up into critical points that lie very near the original one and far from the others.

The sum of all n_i is now n. We again use ℓ for the number of U_i, and write Λ and L for the direct sum of all the Λ_i and the product of all the L_i respectively. For $x \in L$, write $g_x = f + \sum x_{i,j} f_{i,j}$. It remains to show that the set of x for which g_x is not a Morse function—that is, has a critical point with zero Hessian or has two critical points with the same value—is contained in a union of finitely many semialgebraic sets.

Denote elements of W_i by $\overline{u_i} = (u_{i,1}, \ldots, u_{i,m})$, and similarly for elements $\overline{x_i}$ of L_i. Define $G_i \colon W_i \times L_i \to \mathbb{R}$ by $G_i(\overline{u_i}, \overline{x_i}) = p_i(\overline{u_i}) + \sum_{j=1}^{n_i} x_{i,j} q_{i,j}(\overline{u_i})$. Note that for $x = (\overline{x_1}, \ldots, \overline{x_\ell})$, $(G_i)_{\overline{x_i}}$ is exactly the restriction of g_x to W_i.

We introduce one more notation that will be convenient. For $X \subseteq L_i$ define $E(X)$ to be the set of all $(\overline{x_1}, \ldots, \overline{x_\ell})$ in L such that $\overline{x_i} \in X$. When X is a semialgebraic subset of L_i, $E(X)$ is a semialgebraic subset of L. Similarly, if $X \times Y \subseteq L_i \times L_j$, we use $E(X \times Y)$ to denote its extension to a subset of L, that is, $E(X) \cap E(Y)$.

For each i, let S_i be the set of all $(\overline{u_i}, \overline{x_i})$ in $W_i \times L_i$ such that $\partial G_i / \partial u_{i,j}$ for $1 \le j \le n_i$ all vanish at $(\overline{u_i}, \overline{x_i})$, that is, the pairs such that $\overline{u_i}$ is a critical point of $(G_i)_{\overline{x_i}}$. Since S_i is the intersection of an algebraic set in $\mathbb{R}^m \times \mathbb{R}^{n_i}$ with $W_i \times L_i$, and the latter are round open balls, S_i is semialgebraic. Let H_i be the set of all $(\overline{u_i}, \overline{x_i})$ in $W_i \times L_i$ such that the Hessian of $(G_i)_{\overline{x_i}}$ vanishes at $\overline{u_i}$, again a semialgebraic set. The intersection $H_i \cap S_i$ is the set of all $(\overline{u_i}, \overline{x_i})$ such that $(G_i)_{\overline{x_i}}$ has an unstable critical point at $\overline{u_i}$. By the Tarski–Seidenberg Theorem, its projection to L_i is a semialgebraic set, which we will denote by X_i. The union of the $E(X_i)$, $1 \le i \le \ell$, is precisely the set of x in L such that g_x has an unstable critical point.

Now consider $G_i \times G_i \colon S_i \times S_i - \Delta_i \to \mathbb{R}^2$, where Δ_i is the diagonal in $S_i \times S_i$. Let $\widetilde{Y_i} = (G_i \times G_i)^{-1}(\Delta_{\mathbb{R}^2})$, where $\Delta_{\mathbb{R}^2}$ is the diagonal of $\mathbb{R}^2$. Now, let Δ_i' be the set of all $((\overline{u_i}, \overline{x_i}), (\overline{u_i}', \overline{x_i}'))$ in $W_i \times L_i \times W_i \times L_i$ such that $\overline{x_i} = \overline{x_i}'$. Then the projection of $\widetilde{Y_i} \cap \Delta_i'$ to its first two coordinates is the set of all $(\overline{u_i}, \overline{x_i})$ in $W_i \times L_i$ such that $\overline{u_i}$ is a critical point of $(G_i)_{\overline{x_i}}$ and $(G_i)_{\overline{x_i}}$ has another critical point with the same value. The projection to the second coordinate alone is the set Y_i of $\overline{x_i}$ for which $(G_i)_{\overline{x_i}}$ has two critical points with the same value.

Finally, for $i \ne j$, consider $G_i \times G_j \colon S_i \times S_j \to \mathbb{R}^2$ and let $\widetilde{Y_{i,j}}$ be the inverse image of $\Delta_{\mathbb{R}^2}$. Let $Y_{i,j}$ be the projection of $\widetilde{Y_{i,j}}$ to a subset of $L_i \times L_j$. The union of the $E(Y_i)$ and the $E(Y_{i,j})$ is precisely the set of all x such that g_x has two critical points with the same value. Since these are semialgebraic sets, the proof is complete. □

Here is the main result of this subsection.

Theorem 5.7. *Let M and W be compact smooth manifolds. Then for a residual set of smooth maps F from $I \times W$ to $C^\infty(M, \mathbb{R})$, the following hold.*

(i) $F(I \times W) \subset \mathscr{F}_{\ge 0}$.
(ii) *Each $F_u^{-1}(\mathscr{F}_{>0})$ is finite.*

Proof. Start with a smooth map $G \colon I \times W \to C^\infty(M, \mathbb{R})$. Regarding it as a parameterized family of maps $M \times (I \times W) \to \mathbb{R}$, we apply Theorem 5.5 to perturb G so that each G_u is weakly transverse to the points of $\mathbb{R}$. By Proposition 5.7, this implies that $G(I \times W) \subset \mathscr{F}_{\ge 0}$. Since $I \times W$ is compact, $G(I \times W) \subset \mathscr{F}_{\le n}$ for some n.

For each $f \in \mathscr{F}_{>0}$, choose a neighborhood $V_f = U_f \times \mathbb{R}^n$ as in Theorem 5.6. Using Proposition 5.8, we may select a neighborhood L_f of 0 in $\mathbb{R}^n$ and a nonzero polynomial $p_f \colon L_f \to \mathbb{R}$ such that $p_f(L \cap F_{i \ge 1}) = 0$.

Now, partition I into subintervals and triangulate W so that for each subinterval J and each simplex Δ of maximal dimension in the triangulation, $G(J \times \Delta)$ lies either in $\mathscr{F}_0$ or in some $U_f \times L_f$. Fix a particular $J \times \Delta$. If $G(J \times \Delta)$ lies in $\mathscr{F}_0$, do nothing. If not, choose f so that $G(J \times \Delta)$ lies in $U_f \times L_f$. Let $\pi: U_f \times L_f \to L_f$ be the projection, so that $p_f \circ \pi(U_f \times F_{\geq 1}) = 0$. By Proposition 5.5, we may perturb $G|_{J\times\Delta}$ (changing only its L_f-coordinate in $U_f \times L_f$) so that for each $u \in \Delta$, $G_u|_J^{-1}(\mathscr{F}_{i\geq 1})$ is finite, and any map sufficiently close to $G|_{J\times\Delta}$ on $J \times \Delta$ will have this same property. As usual, of course, this is extended to a perturbation of G.

This process can be repeated sequentially on the remaining $J \times \Delta$. The perturbations must be so small that the property of having each $G_u|_J^{-1}(\mathscr{F}_{i\geq 1})$ finite is not lost on previously considered sets. When all $J \times \Delta$ have been considered, each $G_u^{-1}(\mathscr{F}_{i\geq 1})$ is finite. □

5.8.6 Border Label Control

We now return to the case when M is a closed three-manifold, as in the introduction of Sect. 5.8. In this subsection, we will obtain a deformation of $f: M \times W \to M$ for which some ϵ gives border label control.

We begin by ensuring that no f_u carries a component of the singular set T of τ into T. Consider two circles C_1 and C_2 embedded in M. By Theorem 5.4, applied with $A = C_1 \times W$, $B = M$, and $C = C_2$, we may perturb $f|_{C_1\times W}$ so that for each $u \in W$, $f_u|_{C_1}$ meets C_2 in only finitely many points.

Recall that T consists of smooth circles and arcs in M. Each arc is part of some smoothly embedded circle, so T is contained in a union $\cup_{i=1}^n C_i$ of embedded circles in M. By a sequence of perturbations as above, we may assume that at each u, each $f_u(C_i)$ meets each C_j in a finite set (including when $i = j$), so that $f_u(T)$ meets T in a finite set.

The next potential problem is that at some u, $f_u(T_0)$ or $f_u(T_1)$ might be contained in a single level P_t. Recall that the notation $R(s,t)$, introduced in Sect. 5.5, means $\tau^{-1}([s,t])$. For some $\delta > 0$, every $f_u(T_0)$ meets $R(3\delta, 1-3\delta)$, since otherwise the compactness of W would lead to a parameter u for which $f_u(T_0) \subset T$. Let $\phi: R(\delta, 1-\delta) \to [\delta, 1-\delta]$ be the restriction of the map $\pi(\tau(x,t)) = t$. This ϕ makes $R(\delta, 1-\delta)$ a bundle with fibers that are level tori. As before, let C_1 be one of the circles whose union contains T. Only the most superficial changes are needed to the proof of Theorem 5.5 given in [8] so that it applies when ϕ is a bundle map defined on a codimension-zero submanifold of B rather than on all of B; the only difference is that the subsets of jets which are to be avoided are defined only at points of the subspace rather than at every point of B. Using this slight generalization of Theorem 5.5 (and as usual, the Parameterized Extension Principle), we perturb f so that each $f_u|_{C_1}$ is weakly transverse to each P_t with $\delta \leq t \leq 1-\delta$. Since C_1 is one-dimensional, weakly transverse implies that $f_u(C_1)$ meets each such P_t in only finitely many points. Repeating for the other C_i, we may assume that each $f_u(T_0)$ meets the P_t with $\delta \leq t \leq 1-\delta$ in only finitely many points. We also choose the perturbations small enough so that each $f_u(T_0)$ still meets $R(2\delta, 1-2\delta)$. So $f_u^{-1}(P_t) \cap T_0$ is nonempty and finite at least some t. In particular, $\pi(f_u(T_0))$ contains an open set, so by Sard's Theorem applied to $\pi \circ f_u|_{T_0}$, for each u, there is an r so that $f_u(T_0)$ meets P_r transversely in a nonempty set (we select r so that P_r does not contain the image of a vertex of T_0). For a small enough ϵ, a component of $X_s \cap P_r$ will be a compressing disk of X_s whenever $s \leq 2\epsilon$, and by compactness of W, there is an ϵ such that for every u, there is a level P_r such that some component of $X_s \cap P_r$ contains a compressing disk of X_s whenever $s \leq 2\epsilon$.

Applying the same procedure to T_1, we may assume that for every u, there is a level P_r such that some component of $Y_s \cap P_r$ is a compressing disk of Y_s whenever $s \geq 1-2\epsilon$.

Let $h: M \times W \to M$ be defined by $h(x,u) = f_u^{-1}(x)$. Fix new sweepouts on the $M \times \{u\}$, given by $f_u \circ \tau$, so that h_u carries the levels of this sweepout to the original P_t. Applying the previous procedure to h, making sure that all perturbations are small enough to preserve the conditions developed for f, and perhaps making ϵ smaller, we may assume that for each u, there is a level Q_r such that $V_t \cap Q_r$ is a compressing disk of V_t whenever $t \leq 2\epsilon$, and a similar Q_r for W_t with $t \geq 1-2\epsilon$. Thus the number ϵ gives border label control for f. Since border label control holds, with the same ϵ, for any map sufficiently close to f, we may assume it is preserved by all future perturbations.

5.8.7 Building the Graphics

It remains to deform f to satisfy conditions (GP1), (GP2), and (GP3). As before, let $i: I \to \mathbb{R}$ be the inclusion, and consider the smooth map $i \circ \pi \circ f \circ (\tau \times 1_W): P \times \mathrm{I} \times W \to \mathbb{R}$. Regard this as a family of maps from I to $C^\infty(P, \mathbb{R})$, parameterized by W. Apply Theorem 5.7 to obtain a family $k: P \times \mathrm{I} \times W \to \mathbb{R}$. For each $I \times \{u\}$, there will be only finitely many values of s in I for which the restriction $k_{(s,u)}$ of k to $P \times \{s\} \times \{u\}$ is not a Morse function. At these levels, the projection from Q_s into the transverse direction to P_t is an element of some $\mathscr{F}_n$, so each tangency of Q_s with P_t looks like the graph of a critical point of finite multiplicity. This will ultimately ensure that condition (GP1) is attained when we complete our deformations of f.

We will use k to obtain a deformation of the original f, by moving image points vertically with respect to the levels of the range. This would not make sense where the values of k fall outside $(0,1)$, so the motion will be tapered off so as not to change f at points that map near T. It also would not be well-defined at points of $T \times W$, so we taper off the deformation so as not to change f near $T \times W$. The fact that f is unchanged near $T \times W$ and near points that map to T will not matter, since border label control will allow us to ignore these regions in our later work.

Regard $P \times \mathrm{I} \times W$ as a subspace of $P \times \mathbb{R} \times W$. For each $(x, r, u) \in P \times \mathrm{I} \times W$, let $w'_{(x,r,u)}$ be $k(x, r, u) - i \circ \pi \circ f_u \circ \tau(x, r)$, regarded as a tangent vector to $\mathbb{R}$ at $i \circ \pi \circ f_u \circ \tau(x, r)$.

We will taper off the $w'_{(x,r,u)}$ so that for each fixed u they will produce a vector field on M. Fix a number ϵ that gives border label control for f, and a smooth function $\mu: \mathbb{R} \to I$ which carries $(-\infty, \epsilon/4] \cup [1-\epsilon/4, \infty)$ to 0 and carries $[\epsilon/2, 1-\epsilon/2]$ to 1. Define $w_{(x,r,u)}$ to be $\mu(r)\, \mu(i \circ \pi \circ f_u \circ \tau(x, r))\, w'_{(x,r,u)}$. These vectors vanish whenever $r \notin [\epsilon/4, 1 - \epsilon/4]$ or $i \circ \pi \circ f_u \circ \tau(x, r, u) \notin [\epsilon/4, 1 - \epsilon/4]$, that is, whenever $\tau(x, r)$ or $f_u \circ \tau(x, r)$ is close to T. Using the map $i \circ \pi: M \to \mathbb{R}$, we pull the $w_{(x,r,u)}$ back to vectors in M that are perpendicular to P_t; this makes sense near T since the $w_{(x,r,u)}$ are zero at these points). For each u, we obtain at each point $f_u \circ \tau(x, r) \in M$ a vector $v_{(x,r,u)}$ that points in the I-direction (i.e. is perpendicular to P_t) and maps to $w_{(x,r,u)}$ under $(i \circ \pi)_*$.

If k was a sufficiently small perturbation, the $v_{(x,r,u)}$ define a smooth map $j_u: M \to M$ by $j_u(\tau(x, r)) = \mathrm{Exp}(v_{(x,r,u)})$. Put $g_u = j_u \circ f_u$. Since $\mu(r) = 1$ for $\epsilon/2 \le r \le 1 - \epsilon/2$, we have $i \circ \pi \circ g_u \circ \tau(x, r) = k(x, r, u)$ whenever both $\epsilon/2 \le r \le 1 - \epsilon/2$ and $\epsilon/2 \le i \circ \pi \circ f_u \circ \tau(x, r) \le 1 - \epsilon/2$. The latter condition says that $f_u \circ \tau(x, r)$ is in P_s for some $\epsilon/2 \le s \le 1 - \epsilon/2$. Assuming that k was close enough to $i \circ \pi \circ f \circ (\tau \times 1_W)$ so that each $\pi \circ g_u \circ \tau(x, r)$ is within $\epsilon/4$ of $\pi \circ f_u \circ \tau(x, r)$, the equality $i \circ \pi \circ g_u \circ \tau(x, r) = k(x, r, u)$ holds whenever $\tau(x, r)$ is in a P_s and $g_u \circ \tau(x, r)$ is in a P_t with $\epsilon \le s, t \le 1 - \epsilon$.

Carrying out this construction for a sequence of k that converge to $i \circ \pi \circ f \circ (\tau \times 1_W)$, we obtain vector fields $v_{(x,r,u)}$ that converge to the zero vector field. For those sufficiently close to zero, g will be a deformation of f. Choosing g sufficiently close to f, we may ensure that ϵ still gives border label control for g.

We will now analyze the graphic of g_u on I_ϵ^2. For $s, t \in [\epsilon, 1 - \epsilon]$, $\pi \circ g_u(x)$ equals $k_{(s,u)}(x)$ whenever $x \in P_s$ and $g_u(x) \in P_t$. Therefore the tangencies of $g_u(P_s)$ with P_t are locally just the graphs of critical points of $k_{(s,u)}: P \to \mathbb{R}$, so g has property (GP1).

Let $s_1, \ldots, s_n$, be the values of s in $[\epsilon, 1 - \epsilon]$ for which $k_{(s_i,u)}: P \to \mathbb{R}$ is not a Morse function. Each $k_{(s_i,u)}$ is still a function of finite codimension, so has finitely many critical points. Those critical points having critical values in $[\epsilon, 1-\epsilon]$ produce the points of the graphic of g_u that lie in the vertical line $s = s_i$, as suggested in Fig. 5.8. We declare the (s_i, t) at which $k_{(s_i,u)}$ has a critical point at t to be vertices of Γ_u.

When s is not one of the s_i, $k_{(s,u)}$ is a Morse function, so any tangency of $g_u(P_s)$ with P_t is stable, and there is at most one such tangency. Since these tangencies are stable, all nearby tangencies are equivalent to them and hence also stable, so in the graphic for g_u in I_ϵ^2, the pairs (s, t) corresponding to levels with a single stable tangency form ascending and descending arcs as suggested in Fig. 5.8. These arcs may enter or leave I_ϵ^2, or may end at a point corresponding to one of the finitely many points of the graphic with s-coordinate equal to one of the s_i. We declare the intersection points of these arcs with ∂I_ϵ to be vertices of Γ_u. The conditions (GP2) and (GP3) have been achieved, completing the proof of Theorem 5.3.

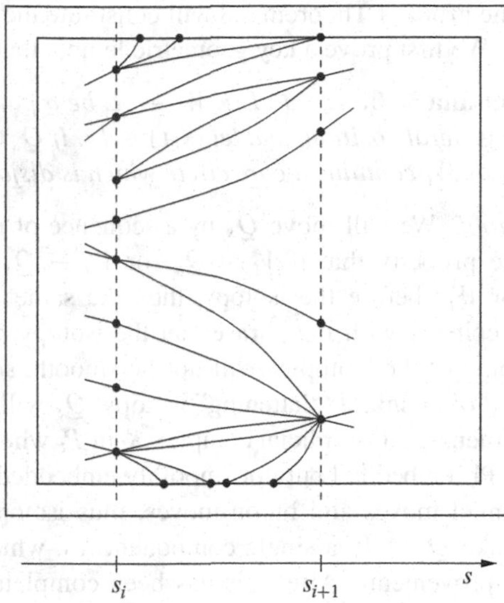

Fig. 5.8 A portion of the graphic of g_u

5.9 Finding Good Regions

In this section, we adapt the arguments of Sect. 5.6 to general position families. The graphics associated to the f_u of a general position family $f: M \times W \to M$ satisfy property (RS1) of Sect. 5.6 (provided that the Heegaard splittings associated to the sweepout are strongly irreducible) and property (RS2) (since the open edges of the Γ correspond to pairs of levels that have a single stable tangency, see the remark after the definition of (RS2) in Sect. 5.6), but not property (RS3). Indeed, property (RS3) does not even make sense, since the vertices of Γ can have high valence. Property (RS1) is what allows the map from the 0-cells of K to the 0-simplices of the Diagram to be defined. Property (RS2) (plus conditions on regions near ∂K, which we will still have due to border label control) allows it to be extended to a cellular map from the 1-skeleton of K to the 1-skeleton of the Diagram. What ensures that it still extends to the 2-cells is a topological fact about pairs of levels whose intersection contains a common spine, Lemma 5.10. Because it involves surfaces that do not meet transversely, its proof is complicated and somewhat delicate. Since the proof does not introduce any ideas needed elsewhere, the reader may wish to skip it on a first reading, and go directly from the statement of Lemma 5.10 to the last four paragraphs of the section.

We specialize to the case of a parameterized family $f: L \times W \to L$, where L is a lens space and W is a compact manifold. We retain the notations P_t, Q_s, V_t, W_t, X_s, and Y_s of Sect. 5.8. As was mentioned above, properties (RS1) and (RS2) already hold for the labels of the regions of the graphic of each f_u.

Theorem 5.8. *Suppose that $f: L \times W \to L$ is in general position with respect to τ. Then for each u, there exists (s,t) such that Q_s meets P_t in good position.*

The proof of Theorem 5.8 will constitute the remainder of this section.

We first prove a key geometric lemma that is particular to lens spaces.

Lemma 5.10. *Let $f\colon L \times W \to L$ be a parameterized family of diffeomorphisms in general position, and let $(s,t) \in I^2_\epsilon$. If $Q_s \cap P_t$ contains a spine of P_t, then either V_t or W_t contains a core circle which is disjoint from Q_s.*

Proof. We will move Q_s by a sequence of isotopies of L. All isotopies will have the property that if $V_t - Q_s$ (or $W_t - Q_s$) did not contain a core circle of V_t (or W_t) before the isotopy, then the same is true after the isotopy. We say this succinctly with the phrase that the isotopy *does not create core circles.* Typically some of the isotopies will not be smooth, so we work in the PL category. At the end of an initial "flattening" isotopy, Q_s will intersect P_t nontransversely in a two-dimensional simplicial complex X in P_t whose frontier consists of points where Q_s is PL embedded but not smoothly embedded. A sequence of simplifications called tunnel moves and bigon moves, plus isotopies that push disks across balls, will make $Q_s \cap P_t$ a single component X_0, which will then undergo a few additional improvements. After this has been completed, an Euler characteristic calculation will show that a core circle disjoint from the repositioned Q_s exists in either V_t or W_t, and consequently one existed for the original Q_s.

The first step is to perform a so-called "flattening" isotopy. Such isotopies were already described in detail in Lemma 4.4, but we will give a self-contained construction here.

Since f is in general position, $Q_s \cap P_t$ is a 1-complex satisfying the property (GP1) of Sect. 5.8. Each isolated vertex of $Q_s \cap P_t$ is an isolated tangency of $Q_s \cap P_t$, so we can move Q_s by a small isotopy near the vertex to eliminate it from the intersection. After this step, $Q_s \cap P_t$ is a graph Γ which contains a spine of $Q_s \cap P_t$, such that each vertex of Γ has positive valence.

By property (GP1), each vertex x of Γ is a point where Q_s is tangent to P_t, and the edges of Γ that emanate from x are arcs where Q_s intersects P_t transversely. Along each arc, Q_s crosses from V_t into W_t or vice versa, so there is an even number of these arcs. Near x, the tangent planes of Q_s are nearly parallel to those of P_t, and there is an isotopy that moves a small disk neighborhood of x in Q_s until it coincides with a small disk neighborhood of x in P_t. Perform such isotopies near each vertex of Γ. This enlarges Γ in $Q_s \cap P_t$ to the union of Γ with a union E of disks, each disk containing one of the original vertices.

The closure of the portion of Γ that is not in E now consists of a collection of arcs and circles where Q_s intersects P_t transversely, except at the endpoints of the arcs, which lie in E. Consider one of these arcs, α. At points of α near E, the tangent planes to Q_s are nearly parallel to those of P_t, and starting from each end there is an isotopy that moves a small regular neighborhood of a portion of α in Q_s onto a small regular neighborhood of the same portion of α in P_t. This flattening process can be continued along α. If it is started from both ends of α, it may be possible to flatten all of a regular neighborhood of α in Q_s onto one in P_t. This occurs when the vectors in a field of normal vectors to α in Q_s are being moved to normal vectors on the same side of α in P_t. If they are being moved to opposite sides, then we

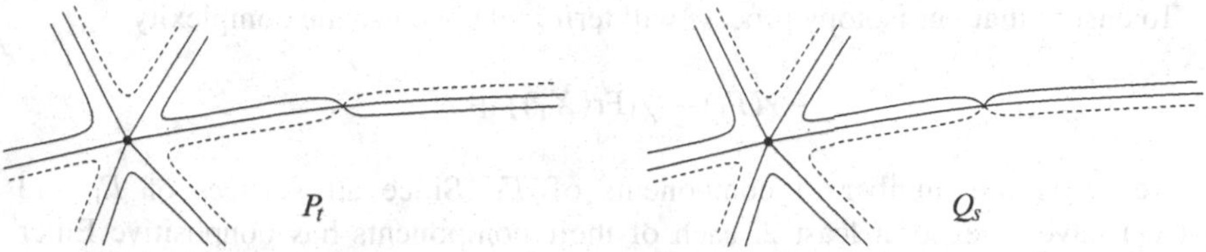

Fig. 5.9 *Up* and *down edges* of X as they appear in P_t and Q_s

introduce a point where the configuration is as in Fig. 4.1, in which P_t appears as the xy-plane, α appears as the points in P_t with $x = -y$, and Q_s appears as the four shaded half- or quarter-planes. These points will be called *crossover* points. Perform such isotopies in disjoint neighborhoods of all the arcs of $\Gamma - E$. For the components of Γ that are circles of transverse intersection points, we flatten Q_s near each circle to enlarge the intersection component to an annulus.

At the end of this initial process, Γ has been enlarged to a 2-complex X in $Q_s \cap P_t$ that is a regular neighborhood of Γ, except at the crossover points where Γ and X look locally like the antidiagonal $x = -y$ of the xy-plane and the set of points with $xy \leq 0$. We will refer to X as a *pinched regular neighborhood* of Γ.

Since Γ originally contained a spine of P_t, X contains two circles that meet transversely in one point that lies in the interior (in P_t) of X. Therefore X contains a common spine of P_t and Q_s. Let X_0 be the component of X that contains a common spine of Q_s and P_t. All components of $P_t - X_0$ and $Q_s - X_0$ are open disks. Let $X_1 = X - X_0$, and for each i, denote $\Gamma \cap X_i$ by Γ_i.

The next step will be to move Q_s by isotopy to remove X_1 from $Q_s \cap P_t$. These isotopies will be fixed near X_0. Some of them will have the effect of joining two components of $V_t - Q_s$ (or of $W_t - Q_s$) into a single component of $V_t - Q_s$ (or of $W_t - Q_s$) for the repositioned Q_s, so we must be very careful not to create core circles.

The frontier of X_1 in P_t is a graph $\mathrm{Fr}(X_1)$ for which each vertex is a crossover point, and has valence 4 (as usual, our "graphs" can have open edges that are circles). Its edges are of two types: *up* edges, for which the component of $\overline{Q_s - X}$ that contains the edge lies in W_t, and *down* edges, for which it lies in V_t. At each disk of E, the up and down edges alternate as one moves around ∂E (see Fig. 5.9). For each of the arcs of $\Gamma_1 - E$, the flattening process creates an up edge on one side and a down edge on the other, but there is a fundamental difference in the way that the up and down edges appear in Q_s and P_t. As shown in Fig. 5.9, up edges (the solid ones) and down edges (the dotted ones) alternate as one moves around a crossover point, while in Q_s they occur in adjacent pairs. This is immediate upon examination of Fig. 4.1.

For our inductive procedure, we start with a pinched regular neighborhood $X_1 \subset Q_s \cap P_t$ of a graph Γ_1 in $Q_s \cap P_t$, all of whose vertices have positive even valence. Moreover, the edges of the frontier of X_1 are up or down according to whether the portion of $\overline{Q_s - X}$ that contains them lies in W_t or V_t. We call this an *inductive configuration.*

To ensure that our isotopy process will terminate, we use the complexity

$$-\chi(\Gamma_1) - \chi(\mathrm{Fr}(X_1)) + N$$

where N is the number of components of Γ_1. Since all vertices of Γ_1 and $\mathrm{Fr}(X_1)$ have valence at least 2, each of their components has nonpositive Euler characteristic, so the complexity is a non-negative integer. The remaining isotopies will reduce this complexity, so our procedure must terminate.

We may assume that the complexity is nonzero, since if $N = 0$ then X_1 is empty. Consider X_1 as a subset of the union of open disks $Q_s - X_0$. Since X_1 is a pinched regular neighborhood of a graph with vertices of valence at least 2, it separates these disks, and we can find a closed disk D in Q_s with $\partial D \subset X_1$ and $D \cap X = \partial D$. It lies either in V_t or W_t. Assume it is in W_t (the case of V_t is similar), in which case all of its edges are up edges. Since $\partial D \subset P_t - X_0$, ∂D bounds a disk D' in $P_t - X_0$. Since the interior of D is disjoint from P_t, $D \cup D'$ bounds a 3-ball Σ in L. Of course, D' may contain portions of the component of X_1 that contains $\partial D'$, or other components of X_1. Let X_1' be the component of X_1 that contains $\partial D'$; it is a pinched regular neighborhood of a component Γ_1' of Γ.

Suppose that X_1' contains some vertices of Γ_1 of valence more than 2. We will perform an isotopy of Q_s that we call a *tunnel move,* illustrated in Fig. 5.10, that reduces the complexity of the inductive configuration. Near the vertex, select an arc in X_1' that connects the edge of $\mathrm{Fr}(X_1')$ in D' with another up edge of $\mathrm{Fr}(X_1')$ that lies near the vertex (this arc may lie in D', in a portion of X_1 contained in D'). An isotopy of Q_s is performed near this arc, that pulls an open regular neighborhood of the arc in X_1' into W_t. This does not change the interior of $V_t - Q_s$ (it just adds the regular neighborhood of the arc to $V_t - Q_s$), but in W_t it creates a tunnel that joins two different components of $W_t - Q_s$. One of these components was in Σ, so the isotopy cannot create core circles. After the tunnel move, we have a new inductive configuration. The Euler characteristic of Γ_1 has been increased by the addition of one vertex, while $\chi(\mathrm{Fr}(X_1))$ and N are unchanged, so the new inductive configuration is of lower complexity. The procedure continues by finding a new D and D' and repeating the process.

When a D has been found for which no tunnel moves are possible, all vertices of Γ_1' (if any) have valence 2. Suppose that X_1' contains crossover points. It must contain an even number of them, since up and down edges at crossover points alternate in P_t. Some portion of X_1' is a disk B whose frontier consists of two crossover points and two edges of $\mathrm{Fr}(X_1)$, each connecting the two crossover points. We will use a *bigon move* as in the proof of Proposition 4.1. A bigon move is an isotopy of Q_s, supported in a neighborhood of B, that repositions Q_s and replaces a neighborhood of B in X with a rectangle containing no crossover points. Figure 4.4 illustrates this isotopy. It cannot create core circles, indeed such an isotopy changes the interiors of $V_t - Q_s$ and $W_t - Q_s$ only by homeomorphism.

Since bigon moves increase the Euler characteristic of $\mathrm{Fr}(X_1)$, without changing Γ_1 or N, they reduce complexity. So we eventually arrive at the case when X_1' is an annulus. Assume for now that the interior of D' is disjoint from X_1. There is an

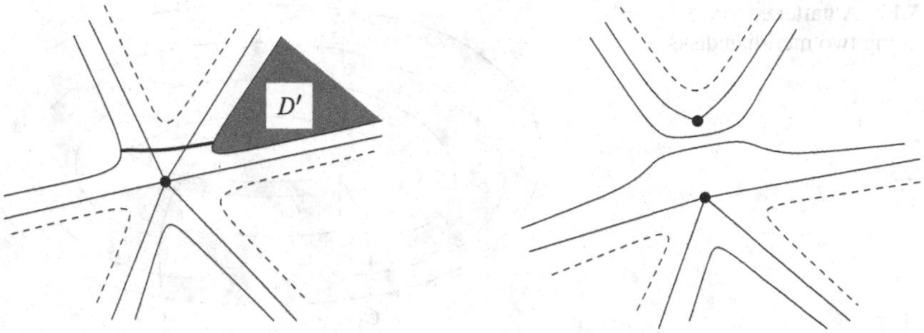

Fig. 5.10 A portion of P_t showing a tunnel arc in X_1, and the new Γ_1 and X_1 after the tunnel move

isotopy of Q_s that pushes D across Σ, until it coincides with D'. This cannot create core circles, since its effect on the homeomorphism type of $W_t - Q_s$ is simply to remove the component $\Sigma - Q_s$. Perform a small isotopy that pulls D' off into the interior of V_t, again creating no new core circles. An annulus component of X_1 has been eliminated, reducing the complexity. If $D' \cap X_1 = X_1'$, then a similar isotopy eliminates X_1'.

Suppose now that the interior of D' contains components of X_1 other than perhaps X_1'. Let X_1'' be their union. It is a pinched regular neighborhood of a union Γ_1'' of components of Γ_1. If Γ_1'' has vertices of valence more than 2, then tunnel moves can be performed. These cannot create new core circles, since they do not change the interior of $V_t - Q_s$, and in $W_t - Q_s$ they only connect regions that are contained in Σ. If no tunnel move is possible, but there are crossover points, then a bigon move may be performed. So we may assume that every component of X_1'' is an annulus.

Let S be a boundary circle of X_1'' innermost on D', bounding a disk D'' in D' whose interior is disjoint from X. Let E'' be the disk in Q_s bounded by S, so that $D'' \cup E''$ bounds a 3-ball Σ'' in L. Note that E'' does not contain X_0, since then a spine of P_t would be contained in the 2-sphere $E'' \cup D''$.

We claim that if $(V_t - Q_s) \cup (E'' \cap V_t)$ contains a core circle of V_t, then $V_t - Q_s$ contained a core circle of V_t (and analogously for W_t). The closures of the components of $E'' - P_t$ are planar surfaces, each lying either in V_t or W_t. Let F be one of these, lying (say) in V_t. Its boundary circles lie in $P_t - X_0$, so bound disks in P_t. The union of F with these disks is homotopy equivalent to $S^2 \vee (\vee S^1)$ for some possibly empty collection of circles, so a regular neighborhood in V_t of the union of F with these disks is a punctured handlebody $Z(F)$ meeting P_t in a union of disks. Suppose that C is a core circle in V_t that is disjoint from $Q_s - F$. We may assume that C meets $\partial Z(F)$ transversely, so cuts through $Z(F)$ is a collection of arcs. Since $Z(F)$ is handlebody meeting P_t only in disks, there is an isotopy of C that pushes the arcs to the frontier of $Z(F)$ in V_t and across it, removing the intersections of C with F without creating new intersections (since the arcs need only be pushed slightly outside of $Z(F)$). Performing such isotopies for all

Fig. 5.11 A flattened torus containing two meridian disks

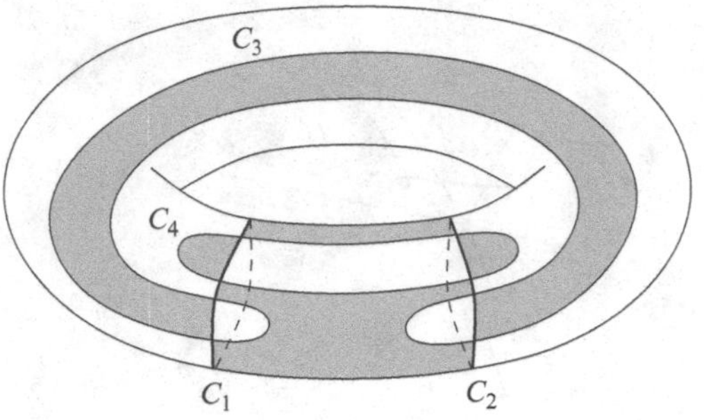

components of $E'' - X_1$ in V_t produces a core circle disjoint from E'', proving the claim.

By virtue of the claim, an isotopy that pushes E'' across Σ'' until it coincides with D'' does not create core circles. Then, a slight additional isotopy pulls D'' and the component of X_1 that contained $\partial D''$ off of P_t, reducing the complexity.

Since we can always reduce a nonzero complexity by one of these isotopies, we may assume that $Q_s \cap P_t = X_0$. The frontier $\mathrm{Fr}(X_0)$ in P_t is the union of a graph Γ_2, each of whose components has vertices of valence 4 corresponding to crossover points, and a graph Γ_3 whose components are circles.

A component of Γ_3 must bound both a disk D_Q in $\overline{Q_s - X_0}$ and a disk D_P in $\overline{P_t - X_0}$. Since $Q_s \cap P_t = X_0$, the interiors of D_P and D_Q are disjoint, and D_Q lies either in V_t or in W_t. So we may push D_Q across the 3-ball bounded by $D_Q \cup D_P$ and onto D_P, without creating core circles. Repeating this procedure to eliminate the other components of Γ_3, we achieve that the frontier of $Q_s \cap P_t$ equals the graph Γ_2.

Figure 5.11 shows a possible intersection of Q_s with P_t at this stage. The shaded region is $Q_s \cap P_t$; it is a union of a (solid) octagon, two bigons, and a square. The closure of $Q_s - (Q_s \cap P_t)$ consists of two meridian disks in V_t, bounded by the circles C_1 and C_2, and two boundary-parallel disks in W_t, bounded by the circles C_3 and C_4.

Suppose that Q_s now contains $2k_1$ meridian disks of Q_s in V_t and $2k_2$ meridian disks in W_t (their numbers must be even since Q_s is zero in $H_2(L)$), and a total of k_0 boundary-parallel disks in V_t and W_t. Since $\chi(Q_s) = 0$, we have $\chi(Q_s \cap P_t) = -k_0 - 2k_1 - 2k_2$. To prove the lemma, we must show that either k_1 or k_2 is 0.

Let V be the number of vertices of Γ_2. Since all of its vertices have valence 4, Γ_2 has $2V$ edges. The remainder of $Q_s \cap P_t$ consists of two-dimensional faces. Each of these faces has boundary consisting of an even number of edges, since up and down edges alternate around a face. If some of the faces are bigons, such as two of the faces in Fig. 5.11, they may be eliminated by bigon moves (which will also change V). These may create additional components of the frontier of X_0 that are circles, indeed this happens in the example of Fig. 5.11. These are eliminated as before by moving disks of Q_s onto disks in P_t. After all bigons have been eliminated, each face has at least four edges, so there are at most $V/2$ faces. So we have $\chi(Q_s \cap P_t) \leq V - 2V + V/2 = -V/2$.

Each boundary-parallel disk in $Q_s \cap V_t$ or $Q_s \cap W_t$ contributes at least two vertices to the graph, since at each crossover point, X_0 crosses over to the other side in P_t of the boundary of the disk. This gives at least $2k_0$ vertices. The meridian disks on the two sides contribute at least $2k_1 \cdot 2k_2 \cdot m$ additional vertices, where $L = L(m, q)$, since the meridians of V_t and W_t have algebraic intersection $\pm m$ in P_t. Thus $V \geq 2k_0 + 4k_1k_2m$. We calculate:

$$-k_0 - 2k_1 - 2k_2 = \chi(Q_s \cap P_t) \leq -V/2 \leq -k_0 - 2k_1k_2m .$$

Since $m > 2$, this can hold only when either k_1 or k_2 is 0. □

Lemma 5.10 fails (at the last sentence of the proof) for the case of $L(2, 1)$. Indeed, there is aflattened Heegaard torus in $L(2, 1)$ which meets $P_{1/2}$ in four squares and has two meridian disks on each side. In a sketch somewhat like that of Fig. 5.11, the boundaries of these disks are two meridian circles and two $(2, 1)$-loops intersecting in a total of eight points, and cutting the torus into eight squares. There are two choices of four of these squares to form $Q_s \cap P_t$.

Now, we will complete the proof of Theorem 5.8. As in Sect. 5.6, assume for contradiction that all regions are labeled, and triangulate I_ϵ^2. The map on the 1-skeleton is defined exactly as in Sect. 5.6, using Lemma 5.6 and the fact that the labels satisfy property (RS2). Using Lemma 5.6, each 1-cell maps either to a 0-simplex or a 1-simplex of the Diagram, and exactly as before the boundary circle of K maps to the Diagram in an essential way. The contradiction will be achieved once we show that the map extends over the 2-cells.

There is no change from before when the 2-cell meets ∂K or lies in the interior of K but does not contain a vertex of Γ, so we fix a 2-cell in the interior of K that is dual to a vertex v_0 of Γ, located at a point (s_0, t_0).

Suppose first that $Q_{s_0} \cap P_{t_0}$ contains a spine of P_{t_0}. By Lemma 5.10, either V_{t_0} or W_{t_0} has a core circle C which is disjoint from Q_{s_0}; we assume it lies in V_{t_0}, with the case when it lies in W_{t_0} being similar. The letter A cannot appear in the label of any region whose closure contains v_0, since C is a core circle for all P_t with t near t_0, and Q_s is disjoint from C for all s near s_0. By Lemma 5.5, any letter a that appears in the label of one of the regions whose closure contains v_0 must appear in a combination of either ax or ay, so none of these regions has label A. Since each 1-cell maps to a 0- or 1-simplex of the Diagram, the map defined on the 1-cells of K maps the boundary of the 2-cell dual to v_0 into the complement of the vertex A of the Diagram. Since this complement is contractible, the map can be extended over the 2-cell.

Suppose now that $Q_{s_0} \cap P_{t_0}$ does not contain a spine of P_{t_0}. Then there is a loop $C_{(s_0,t_0)}$ essential in P_{t_0} and disjoint from Q_{s_0}. For every (s, t) near (s_0, t_0), there is a loop $C_{(s,t)}$ essential in P_t and disjoint from Q_s, with the property that $C_{(s,t)}$ is a meridian of V_t (respectively W_t) if and only if $C_{(s_0,t_0)}$ is a meridian of V_{t_0} (respectively W_{t_0}). In particular, any intersection circle of Q_s and P_t which bounds a disk in Q_s which is precompressing for P_t in V_t or in W_t must be disjoint from $C_{(s,t)}$. Since the meridian disks of V_t and W_t have nonzero algebraic intersection, the meridians for V_t and W_t cannot both be disjoint from $C_{(s,t)}$. So for all (s,t) in this neighborhood of (s_0,t_0), either all disks in Q_s that are precompressions for P_t are precompressions in V_t, or all are precompressions in W_t. In the first case, the letter B does not appear in the label of any of the regions whose closure contain v_0, while in the second case, the letter A does not. In either case, the extension to the 2-cell can now be obtained just as in the previous paragraph. This completes the proof of Theorem 5.8.

5.10 From Good to Very Good

By virtue of Theorem 5.8, we may perturb a parameterized family of diffeomorphisms of M so that at each parameter u, some level P_t and some image level $f_u(P_s)$ meet in good position. In this section, we use the methodology of Hatcher [22,23] (see [25] for a more detailed version of [23], see also Ivanov [32]) to change the family so that we may assume that P_t and $f_u(P_s)$ meet in very good position. In fact, we will achieve a rather stronger condition on discal intersections.

Following our usual notation, we fix a sweepout $\tau\colon P \times [0,1] \to M$ of a closed orientable three-manifold M, and give P_t, V_t, and W_t their usual meanings. Given a parameterized family of diffeomorphisms $f\colon M \times W \to M$, we give f_u, Q_s, X_s, and Y_s their usual parameter-dependent meanings. From now on, we refer to the P_t as *levels* and the Q_s as *image levels*.

Throughout this section, we assume that for each $u \in W$, there is a pair (s,t) such that Q_s and P_t are in good position. Before stating the main result, we will need to make some preliminary selections.

By transversality, being in good position is an open condition, so there exist a finite covering of W by open sets U_i, $1 \le i \le n$, and pairs (s_i,t_i), so that for each $u \in U_i$, Q_{s_i} and P_{t_i} meet in good position. By shrinking of the open cover, we can and always will assume that all transversality and good-position conditions that hold at parameters in U_i actually hold on $\overline{U_i}$.

We want to select the sets and parameters so that at parameters in U_i, Q_{s_i} is transverse to P_{t_j} for all t_j. First note that for any s sufficiently close to s_i, Q_s is transverse to P_{t_i} at all parameters of U_i (here we are already using our condition that the transversality for the Q_{s_i} holds for all parameters in $\overline{U_i}$). On U_1, Q_{s_1} is already transverse to P_{t_1}. Sard's Theorem ensures that at each $u \in U_2$, there is a value s arbitrarily close to s_2 such that Q_s is transverse to P_{t_1} at all parameters in a neighborhood of u. Replace U_2 by finitely many open sets (with associated s-values), for which on each of these sets the associated Q_s are transverse to P_{t_1}. The new s are selected close enough to s_2 so that these Q_s still meet P_{t_2} in good position. Repeat this process for U_3, that is, replace U_3 by a collection of sets and associated values of s for which the associated Q_s are transverse to P_{t_1} and still meet P_{t_3} in good position. Proceeding through the remaining original U_i, we have a new collection, with many more sets U_i, but only the same t_i values that we started with, and at each parameter in one of the new U_i, Q_{s_i} is transverse to P_{t_1} as well as to P_{t_i}. Now proceed to P_{t_2}. For the U_i whose associated t-value is not t_2, we perform a similar process, and we also select the new s-values so close to s_i that the new Q_s are still transverse to P_{t_1} and still meet their associated P_{t_i} in good position. After finitely many repetitions, all Q_{s_i} are transverse to each P_{t_j}.

We may also assume the U_i are connected, by making each connected component a U_i. Since transversality is an open condition, we are free to replace s_i by a very nearby value, while still retaining the good position of Q_{s_i} and P_{t_i} and the transverse intersection of Q_{s_i} with all P_{t_j}, for all parameters in U_i, and similarly we may reselect any t_j. So (with the argument in the previous paragraph now completed) we can and always will assume that all s_i are distinct, and all t_i are distinct.

We can now state the main result of this section. With notation as above:

Theorem 5.9. *Let $f: W \to \mathrm{diff}(M)$ be a parameterized family, such that for each u there exists (s,t) such that Q_s and P_t meet in good position. Then f may be changed by homotopy so that there exists a covering $\{U_i\}$ as above, with the property that for all $u \in U_i$, Q_{s_i} and P_{t_i} meet in very good position, and Q_{s_i} has no discal intersection with any P_{t_j}. If these conditions already hold for all parameters in some closed subset W_0 of W, then the deformation of f may be taken to be constant on some neighborhood of W_0.*

Before starting the proof, we introduce a simplifying convention. Although strictly speaking, Q_{s_i} is meaningful at every parameter, as is every Q_s, throughout the remainder of this section we speak of Q_{s_i} only for parameters in $\overline{U_i}$. That is, unless explicitly stated otherwise, an assertion made about Q_{s_i} means that the assertion holds at parameters in $\overline{U_i}$, but not necessarily at other parameters. Also, to refer to Q_{s_i} at a single parameter u, we use the notation $Q_{s_i}(u)$. By our convention, $Q_{s_i}(u)$ is meaningful only when u is a value in $\overline{U_i}$.

Now, to preview some of the complications that appear in the proof of Theorem 5.9, consider the problem of removing, just for a single parameter $u \in U_i$, a discal component c of the intersection of $Q_{s_i}(u)$ with some P_{t_j}. Suppose that the disk D' in $Q_{s_i}(u)$ bounded by c is innermost among all disks in $Q_{s_i}(u)$ bounded by discal intersections of $Q_{s_i}(u)$ with the P_{t_k}. Note that D' can contain a nondiscal intersection of $Q_{s_i}(u)$ with a P_{t_k}; such an intersection will be a meridian of either V_{t_k} or W_{t_k} (although k cannot equal i, since $Q_{s_i}(u)$ and P_{t_i} meet in good position). Let D be the disk in P_{t_j} bounded by c, so that $D \cup D'$ is the boundary of a 3-ball E. There is an isotopy of f_u that moves D' across E to D, and on across D, eliminating c and possibly other intersections of the $Q_{s_\ell}(u)$ with the P_{t_k}. We will refer to this as a *basic isotopy.*

It is possible for a basic isotopy to remove a biessential component of some $Q_{s_k}(u) \cap P_{t_k}$. Examples are a bit complicated to describe, but involve ideas similar to the construction in Fig. 5.2. Fortunately, the following lemma ensures that good position is not lost.

Lemma 5.11. *After a basic isotopy as described above, each $Q_{s_k}(u) \cap P_{t_k}$ still has a biessential component.*

Proof. Throughout the proof of the lemma, Q_s is understood to mean $Q_s(u)$.

Suppose that a biessential component of some $Q_{s_k} \cap P_{t_k}$ is contained in the ball E, and hence is removed by the isotopy. Since a spine of Q_{s_k} cannot be contained in a 3-ball, there must be a circle of intersection of Q_{s_k} with D that is essential in Q_{s_k}. This implies that $k \neq j$. Now D' must have nonempty intersection with P_{t_k}, since otherwise P_{t_k} would be contained in E. An intersection circle innermost on D' cannot be inessential in P_{t_k}, since c was an innermost discal intersection on Q_{s_i}, so D' contains a meridian disk D'_0 for either V_{t_k} or W_{t_k}. Choose notation so that D is contained in V_{t_k} (that is, $t_j < t_k$).

Suppose first that $D'_0 \subset V_{t_k}$. The basic isotopy pushing D' across E moves $Q_{s_k} \cap E$ into a small neighborhood of D, so that it is contained in V_{t_k}. If there is no longer any biessential intersection of Q_{s_k} with P_{t_k}, then the complement in V_{t_k} of the original D'_0 contains a spine of Q_{s_k} (since the original intersection of Q_{s_k} with D contained a loop essential in Q_{s_k}, the spine of Q_{s_k} is now on the V_{t_k}-side of P_{t_k}). This is a contradiction, since Q_{s_k} is a Heegaard torus.

Suppose now that $D'_0 \subset W_{t_k}$. Since the biessential circles of $Q_{s_k} \cap P_{t_k}$ are disjoint from D'_0, they are meridians for W_{t_k} and hence are essential in V_{t_k}. Now, let A be innermost among the annuli on Q_{s_k} bounded by a biessential component C of $Q_{s_k} \cap P_{t_k}$ and a circle of $Q_{s_k} \cap D$. Since Q_{t_k} and P_{t_k} meet in good position, the intersection of the interior of A with P_{t_k} is discal. This implies that C is contractible in V_{t_k}, a contradiction. □

Proof (of Theorem 5.9). We will adapt the approach used by Hatcher [22]. The principal difference for us is that in [22], there is only a single domain level, whereas we have the different Q_{s_i} on the sets U_i.

The first step is to construct a family $h_{u,t}$, $0 \leq t \leq 1$ of isotopies of the $f_u = h_{u,0}$, which eliminates the discal intersections of every $Q_{s_i}(u)$ with every P_{t_j}. Let $\mathscr{C}$ be the set of discal intersection curves of all $Q_{s_i} \cap P_{t_j}$ for all u, where as previously explained, this refers only to the $Q_{s_i}(u)$ with $u \in U_i$. Since Q_{s_i} is transverse to P_{t_j} at all $u \in \overline{U_i}$, the curves in $\mathscr{C}$ fall into finitely many families which vary by isotopy as the parameter u moves over (the connected set) $\overline{U_i}$. Thus we may regard $\mathscr{C}$ as a disjoint union containing finitely many copies of each $\overline{U_i}$. It projects to W, with the inverse image of u consisting of the discal intersection curves $\mathscr{C}_u$ of the $Q_{s_i}(u)$ and P_{t_j} for which $u \in U_i$. By assumption, no element of $\mathscr{C}$ projects to any parameter $u \in W_0$.

Each $c \in \mathscr{C}_u$ bounds unique disks $D_c \subset P_{t_j}$ and $D'_c \subset Q_{s_i}(u)$ for some i and j. The inclusion relations among the D_c define a partial ordering $<_P$ on $\mathscr{C}_u$, by the rule that $c_1 <_P c_2$ when $D_{c_1} \subset D_{c_2}$. Similarly, $c_1 <_Q c_2$ when $D'_{c_1} \subset D'_{c_2}$.

If c is minimal for $<_Q$, then $D'_c \cup D_c$ is an embedded 2-sphere in M which bounds a 3-ball E_c. By Lemma 5.11, the basic isotopy that pushes D'_c across E_c to D_c and on to the other side of D_c retains the property that every $Q_{s_k}(u) \cap P_{s_k}$ has a biessential intersection. This ensures that when all discal intersections have been eliminated, each $Q_{s_k}(u) \cap P_{t_k}$ will still intersect, so they will be in very good position.

Shrink the open cover $\{U_i\}$ to an open cover $\{U'_i\}$ for which each $\overline{U'_i} \subset U_i$. To construct the $h_{u,t}$, Hatcher introduced an auxiliary function $\Psi\colon \mathscr{C} \to (0, 2)$ that gives the order in which the elements of $\mathscr{C}$ are to be eliminated, and allows the basic isotopies to be tapered off as one nears the frontier of U_i. Denoting by ψ_u the restriction of Ψ to $\mathscr{C}_u$, we will select Ψ so that the following conditions are satisfied:

1. $\psi_u(c) < \psi_u(c')$ whenever $c <_Q c'$
2. $\psi_u(c) < 1$ if $c \subset Q_{s_i}(u)$ and $u \in \overline{U_i'}$
3. $\psi_u(c) > 1$ if $c \subset Q_{s_i}(u)$ and $u \in \overline{U_i} - U_i$

One way to construct such a Ψ is to choose a Riemannian metric on $\tau(P \times (0,1))$ for which each P_t has area 1, and define $\Psi_0(c)$ to be the area of $f_u^{-1}(D_c')$ in P_{s_i}. Then, choose continuous functions α_i which are 0 on $\overline{U_i'}$ and 1 on $W - U_i$, and define $\Psi(c) = \Psi_0(c) + \alpha_i(u)$ for $c \subset Q_{s_i}(u)$.

Roughly speaking, the idea of Hatcher's construction is to have $h_{u,t}$ perform the basic isotopy that eliminates c during a small time interval $I_u(c)$ which starts at the number $\psi_u(c)$. In order to retain control of this process, preliminary steps must be taken to ensure that basic isotopies that move points in intersecting 3-balls E_c do not occur at the same time.

If c is a discal intersection of U_{s_i} and P_{t_j}, we denote U_i and U_i' by $U(c)$ and $U'(c)$. For a fixed isotopic family of $c \in \mathscr{C}$ with $c \subset Q_{s_i}$, the points $(u, \psi_u(c))$ form a d-dimensional sheet $i(c)$ lying over $\overline{U(c)}$, where d is the dimension of W. If $i(c_1)$ meets $i(c_2)$, then by the first property of Ψ, c_1 and c_2 cannot be $<_Q$-related.

Thicken each $i(c)$ to a plate $I(c)$ intersecting each $\{u\} \times [0,2]$ in an interval $I_u(c) = [\psi_u(c), \psi_u(c) + \epsilon]$, for some small positive ϵ. This interval will contain the t-support of the portion of $h_{u,t}$ that eliminates c, assuming that all other loops in $\mathscr{C}_u$ with smaller ψ_u-value have already been eliminated. By condition (1), c will be $<_Q$-minimal at the times $t \in I_u(c)$. Since $\mathscr{C}_u$ is empty for $u \in W_0$, the $h_{u,t}$ will be constant for all $u \in W_0$.

Choose the ϵ small enough so that $I(c_1) \cap I(c_2)$ is nonempty only near the intersections of $i(c_1)$ and $i(c_2)$. This ensures that if basic isotopies eliminating c_1 and c_2 occur on overlapping time intervals, then c_1 and c_2 are $<_Q$-unrelated. Also, choose ϵ small enough so that $I_u(c) \subset [0,1]$ whenever $u \in U_i'$.

Write G_0 for the union of the $i(c)$, and G for the union of the $I(c)$.

It may happen that for some $c_1, c_2 \in \mathscr{C}_u$ with $\psi_u(c_1) < \psi_u(c_2)$, we have $c_2 <_P c_1$. In this case the isotopy which eliminates c_1 will also eliminate c_2. So reduce G by deleting all points $(u, \psi_u(c_2))$ such that $\psi_u(c_1) < \psi_u(c_2)$ for some c_1 with $c_2 <_P c_1$. Make a corresponding reduction of $I(c_2)$ by deleting points $t \in I_u(c_2)$ such that $t > \psi_u(c_1)$ for some c_1 with $c_2 <_P c_1$.

There is a subtle danger here. Suppose that in the previous paragraph, $u \in U(c_1) - \overline{(U'(c_1))}$. If part of $I_u(c_1)$ extends into $(1,2]$, then the isotopy eliminating c_1 may not be completed, and therefore c_2 would not be eliminated. If $u \in \overline{U(c_2)} - \overline{U'(c_2)}$ this does not matter, since we only need to complete the elimination of c_2 at parameters in $U'(c_2)$. But the plate thickness ϵ must be selected small enough so that $I_u(c_2)$ lies in $[0,1]$ at all u for which there is a c_2 with $u \in \overline{U'(c_2)}$ and $\psi_u(c_2) > \psi_u(c_1)$. This is possible because the set of such u is a compact subset of U_i.

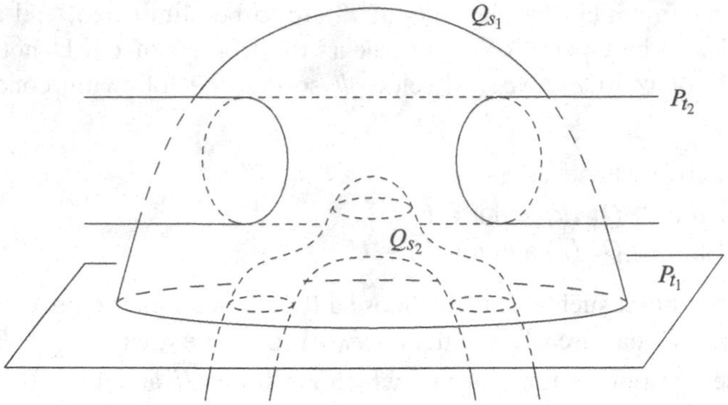

Fig. 5.12 Nested ball regions for basic isotopies

At values of t where the interiors of $I(c_1)$ and $I(c_2)$ still overlap, c_1 and c_2 are $<_Q$-unrelated, and the reduction just made ensures that they are $<_P$-unrelated. In Hatcher's context, all intersections are discal, so the combined effect of these is to eliminate the possibility of simultaneous isotopies on intersecting 3-balls E_{c_1} and E_{c_2}. In our context, however, E_{c_1} and E_{c_2} can intersect on overlaps of $I(c_1)$ and $I(c_2)$ even when c_1 and c_2 are neither $<_P$-related nor $<_Q$-related. Figure 5.12 shows a simple example. The intersections of P_{t_1} with Q_{s_2}, are not discal, nor are the intersections of P_{t_2} with Q_{s_1}, but Q_{s_2} has a discal intersection with P_{t_2} inside $E(c_1)$. When this happens, however, E_{c_1} and E_{c_2} must be either disjoint or nested:

Lemma 5.12. *Suppose that c_1 and c_2 are $<_Q$-minimal discal intersections, and are neither $<_P$-related nor $<_Q$-related. Then ∂E_{c_1} and ∂E_{c_2} are disjoint.*

Proof. Since c_1 and c_2 are not $<_Q$-related, $D'(c_1)$ and $D'(c_2)$ are disjoint, and since they are not $<_P$-related, $D(c_1)$ and $D(c_2)$ are disjoint. An intersection circle of $D(c_1)$ and $D'(c_2)$ would be smaller than c_2 in the $<_Q$-ordering, and similarly an intersection circle of $D'(c_1)$ and $D(c_2)$ would be smaller than c_1 in the $<_Q$-ordering. □

When E_{c_1} and E_{c_2} are nested, say, E_{c_2} lies in E_{c_1}, a basic isotopy that removes c_1 will also remove c_2. So we make the further reduction in G_0 of deleting all $(u, \psi_u(c_2))$ for which there is a c_1 such that $i(c_1)$ meets $i(c_2)$, $\psi_u(c_1) < \psi_u(c_2)$, and $E_{c_2} \subset E_{c_1}$. Also, reduce $I(c_2)$ by removing any t in $I_u(c_2)$ with $t > \psi_u(c_1)$. Again, this may require the plate thickness to be decreased to ensure that $I_u(c_1)$ lies in $[0, 1]$ at parameters in $U(c_1)$ where $u \in \overline{U'(c_2)}$.

For fixed $u \in W$, the basic isotopies are combined by proceeding upward in $W \times [0, 2]$ from $t = 0$ to $t = 1$, performing each basic isotopy involving c on the interval $I_u(c)$. Condition (3) on the ψ_u ensures that the basic isotopies involving $c \subset Q_{s_i}(u)$ taper off at parameters near the frontier of U_i. On a reduced interval $I_u(c)$, which is an initial segment of $[\psi_u(c), \psi_u(c) + \epsilon]$, perform only the corresponding initial portion of the basic isotopy. On the overlaps of the $I(c)$, perform the corresponding basic isotopies concurrently; the reductions of the $I(c)$ have ensured that these basic isotopies will have disjoint supports. Since ϵ was chosen small enough so that $I_u(c) \subset [0, 1]$ whenever $u \in U_i'$, the basic isotopies involving Q_{s_i} will be completed at all u in U_i'. Since $\mathscr{C}_u$ is empty for $u \in W_0$, no isotopies take place at parameters in W_0.

The remaining concern is that the basic isotopies eliminating $c \subset Q_{s_i}(u)$ must be selected so that they fit together continuously in the parameter u on U_i. This can be achieved using the method in the last paragraph on p. 345 of [22] (which applies in the smooth category by virtue of [24], see also the more detailed version in [25]). □

5.11 Setting up the Last Step

In this section, we present some technical lemmas that will be needed for the final stage of the proof.

The first two lemmas give certain uniqueness properties for the fiber of the Hopf fibration on L. Both are false for $\mathbb{RP}^3$, so require our convention that $L = L(m, q)$ with $m > 2$, and as usual we select q so that $1 \le q < m/2$. From now on, we endow L with the Hopf fibering and assume that our sweepout of L is selected so that each P_t is a union of fibers. Consequently the exceptional fibers, if any, will be components of the singular set S.

Lemma 5.13. *Let P be a Heegaard torus in L which is a union of fibers, bounding solid tori V and W. Suppose that a loop in P is a longitude for V and for W. Then $q = 1$ and the loop is isotopic in P to a fiber.*

Proof. Let a and b be loops in P which are respectively a longitude and a meridian of V, and with a determined by the condition that $ma + qb$ is a meridian of W. Let c be a loop in P which is a longitude for both V and W. Since c is a longitude of V, it has (for one of its two orientations) the form $a + kb$ in $H_1(P)$ for some k. The intersection number of c with $ma + qb$ is $q - km$, which must be ± 1 since c is a longitude of W. Since $1 \le q < m/2$ and $m > 2$, this implies that $k = 0$ and $q = 1$. Since $k = 0$, c is uniquely determined and $c = a$. Since $q = 1$, the Hopf fibering is nonsingular, so the fiber is a longitude of both V and W and hence is isotopic in P to c. □

Lemma 5.14. *Let $h: L \to L$ be a diffeomorphism isotopic to the identity, with $h(P_s) = P_t$. Then the image of a fiber of P_s is isotopic in P_t to a fiber.*

Proof. Composing f with a fiber-preserving diffeomorphism of L that moves P_s to P_t, we may assume that $s = t$. Write P, V, and W for P_t, V_t, and W_t. Let a and b be loops in P selected as in the proof of Lemma 5.13, and write $h_*: H_1(P) \to H_1(P)$ for the induced isomorphism.

Suppose first that $h(V) = V$. Since the meridian disk of V is unique up to isotopy, we have $h_*(b) = \pm b$. Since h is isotopic to the identity on L and $m > 2$, h is orientation-preserving and induces the identity on $\pi_1(V)$. This implies that $h_*(b) = b$. Similar considerations for W show that $h_*(ma + qb) = ma + qb$, so $h_*(a) = a$. Thus h_* is the identity on $H_1(P)$ and the lemma follows for this case.

Suppose now that $h(V) = W$. Then h is orientation-reversing on P. Since h must take a meridian of V to one of W, we have $h_*(b) = \epsilon(ma+qb)$ where $\epsilon = \pm 1$. Writing $h_*(a) = ua + vb$, we find that $1 = a \cdot b = -h_*(a) \cdot h_*(b) = -\epsilon(qu - mv)$. The facts that h is isotopic to the identity on L, a generates $\pi_1(L)$, and b is 0 in $\pi_1(V)$ imply that $u \equiv 1 \pmod{m}$, so modulo m we have $1 \equiv -\epsilon q$. Since $1 \le q < m/2$, this forces $q = 1$, $\epsilon = -1$, and $h_*(b) = -ma - b$. Since a has intersection number -1 with the meridian $-ma-b$ of W, it is also a longitude of W. Since h is a homeomorphism interchanging V and W, $h(a)$ is also a longitude of V and of W, and an application of Lemma 5.13 completes the proof. □

We now give several lemmas which allow the deformation of diffeomorphisms and embeddings to make them fiber-preserving or level-preserving. The first is just a special case of Theorem 3.13:

Lemma 5.15. *Let X be either a solid torus or $S^1 \times S^1 \times \mathrm{I}$, with a fixed Seifert fibering. Then the inclusion* $\mathrm{diff}_f(X) \to \mathrm{diff}(X)$ *is a homotopy equivalence.*

Lemma 5.15 guarantees that if $g\colon \Delta \to \mathrm{diff}(X)$ is a continuous map from an n-simplex, $n \ge 1$, with $g(\partial\Delta) \subset \mathrm{diff}_f(X)$, then g is homotopic relative to $\partial\Delta$ to a map with image in $\mathrm{diff}_f(X)$.

The next lemma is a two-dimensional version of Theorem 3.13, and can be proven using surface theory. In fact, it can be proven by applying Theorem 3.13 to $T \times \mathrm{I}$, although that would be a strange way to approach it.

Lemma 5.16. *Let T be a torus with a fixed S^1-fibering. Let* $\mathrm{Diff}_h(T)$ *be the subgroup of* $\mathrm{Diff}(T)$ *consisting of the diffeomorphisms that take some fiber to a loop isotopic to a fiber. Then the inclusion* $\mathrm{Diff}_f(T) \to \mathrm{Diff}_h(T)$ *is a homotopy equivalence.*

For $e \in (0,1)$ we let eD^2 denote the concentric disk of radius e in the standard disk $D^2 \subset \mathbb{R}^2$. Let X be either a solid torus $D^2 \times S^1$, or $T \times \mathrm{I}$ where T is a torus. Let $F = \cup F_i$ be a disjoint union of finitely many tori. Fix an inclusion of F into X such that each F_i is of the form $\partial(e_i D^2 \times S^1)$, in the solid torus case, or of the form $T \times \{e_i\}$, in the $T^2 \times \mathrm{I}$ case, for distinct numbers e_i in $(0,1)$. Let $\mathrm{emb}_{int}(F, X)$ be the connected component of the inclusion in the space of all embeddings of F into the interior of X, and let $\mathrm{emb}_{conc}(F, X)$ be the connected component of the inclusion in the set of embeddings for which each F_i is of the form $\partial(eD^2) \times S^1$ or $T \times \{e\}$ for some $e \in (0,1)$. We omit the proof of the next lemma, which is analogous to Lemma 4.5.

Lemma 5.17. *Let X be a Seifert-fibered solid torus or $S^1 \times S^1 \times \mathrm{I}$. Then the inclusion* $\mathrm{emb}_{conc}(F, X) \to \mathrm{emb}_{int}(F, X)$ *is a homotopy equivalence.*

5.12 Deforming to Fiber-Preserving Families

We are now ready for a key step, deforming a family of diffeomorphisms to preserve the fibering. We remind the reader that this was not possible for the case of $L(4,1)$ in Chap. 4, because neither the longitudinal nor the meridional fibering was preserved by the group of isometries. It is possible now because we are using the Hopf fibering.

Theorem 5.10. *Let* $L = L(m,q)$ *with* $m > 2$ *and let* $f\colon S^d \to \operatorname{diff}(L)$. *Then* f *is homotopic to a map into* $\operatorname{diff}_f(L)$.

Proof. Applying Theorems 5.3, 5.8, and 5.9, we may assume that f satisfies the conclusion of Theorem 5.9. That is, there are pairs (s_i, t_i) and an open cover $\{U_i\}$ of S^d with the property that for every $u \in U_i$, $Q_{s_i}(u)$ and P_{t_i} meet in very good position, and $Q_{s_i}(u)$ meets every P_{t_j} transversely, with no discal intersections. The U_i are selected to be connected, so the intersection $Q_{s_i}(u) \cap P_{t_j}$ is independent, up to isotopy in P_{t_j}, of the parameter u. We remind the reader of our convention that assertions about Q_{s_i} implicitly mean "for every $u \in U_i$." We can and always will assume that conditions stated for parameters in U_i actually hold for all parameters in $\overline{U_i}$.

Since the t_j are distinct, we may select notation so that $t_1 < t_2 < \cdots < t_m$. The corresponding s_i typically are not in ascending order. Figure 5.13 shows a schematic picture of a block of three levels for which the image levels Q_{s_1}, Q_{s_2}, and Q_{s_3} have $s_1 < s_3 < s_2$.

The basic idea of the proof is to make the f_u fiber-preserving on the P_{s_i}, then use Lemma 5.15 to make the f_u fiber-preserving on the complementary $S^1 \times S^1 \times I$ or solid tori of the P_{s_i}-levels. We must be very careful that none of the isotopic adjustments to a Q_{s_i} destroys any condition that must be preserved on the other Q_{s_j}.

Before listing the steps in the proof of Theorem 5.10, a definition is needed. For each i, the intersection circles of $Q_{s_i} \cap P_{t_i}$ cannot be meridians in both V_{t_i} and W_{t_i}, so Q_{s_i} must satisfy exactly one of the following:

1. The circles of $Q_{s_i} \cap P_{t_i}$ are not longitudes or meridians for V_{t_i}, so the annuli of $Q_{s_i} \cap V_{t_i}$ are uniquely boundary parallel in V_{t_i}.
2. The circles of $Q_{s_i} \cap P_{t_i}$ are longitudes or meridians for V_{t_i}, but are not longitudes or meridians for W_{t_i}, so the annuli of $Q_{s_i} \cap W_{t_i}$ are uniquely boundary parallel in W_{t_i}.
3. The circles of $Q_{s_i} \cap P_{t_i}$ are longitudes both for V_{t_i} and for W_{t_i}.

In the first case, we say that Q_{s_i} and P_{t_i} are *V-cored,* in the second that they are *W-cored,* and in the third that they are *bilongitudinal.* If they are either V-cored or W-cored, we say they are *cored.* Lemma 5.13 shows that the bilongitudinal case can occur only when $q = 1$, and then only when the intersection circles are isotopic in P_{t_i} to fibers of the Hopf fibering.

We can now list the steps in the procedure. In this list, and in the ensuing details, "push Q_{s_i}" means perform a deformation of f that moves Q_{s_i} as stated, and preserves all other conditions needed. Making Q_{s_i} "vertical" (at a parameter u)

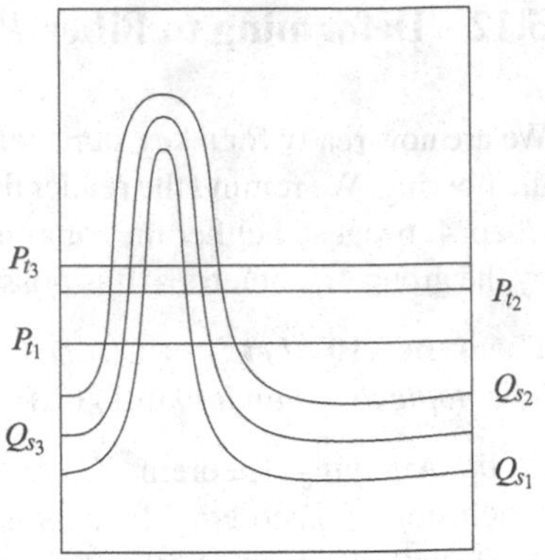

Fig. 5.13 A block of level tori with the Q_{s_i} out of order

means making the restriction of f_u to P_{s_i} fiber-preserving. When we say that something is done "at all parameters of U_i," we mean that a deformation of f will be performed, and that U_i is replaced by a smaller set, so that the result is achieved for all parameters in the new $\overline{U_i}$, while retaining all other needed properties (such as that $\{U_i\}$ is an open covering of S^d).

1. Push the Q_{s_i} that meet P_{t_j} out of V_{t_j}, for all the V-cored P_{t_j}, at all parameters in $U(t_j)$. At the end of this step, each Q_{s_i} that was V-cored is parallel to P_{t_i}.
2. Push the Q_{s_i} that meet P_{t_j} out of W_{t_j}, for all the W-cored P_{t_j}, at all parameters in $U(t_j)$. At the end of this step, each Q_{s_i} that was W-cored is parallel to P_{t_i}.

These first two steps are performed using a method of Hatcher like that of the proof of Sect. 5.10, although simpler. After they are completed, a triangulation of S^d is fixed with mesh smaller than a Lebesgue number for the open cover by the U_i. Each of the remaining steps is performed by inductive procedures that move up the skeleta of the triangulation, achieving the objective for Q_{s_i} at all parameters that lie in a simplex completely contained in U_i.

3. Push the Q_{s_i} that originally were cored so that each one equals some level torus. These level tori may vary from parameter to parameter.
4. Push the Q_{s_i} that originally were cored to be vertical.
5. Push the bilongitudinal Q_{s_i} to be vertical.
6. Use Lemma 5.15 to make f_u fiber-preserving on the complementary $S^1 \times S^1 \times I$ or solid tori of the P_{s_i}-levels.

The underlying fact that allows all of this pushing to be carried out without undoing the results of the previous work is Lemma 5.4. Its use involves the concepts of *compatibility* and *blocks,* which we will now define.

Recall that $R(t_i, t_j)$ means the closure of the region between P_{t_i} and P_{t_j}. For a connected subset Z of S^d, which in practice will be either a single parameter or a simplex of a triangulation, denote by B_Z the set of t_i such that $Z \subset U_i$. Elements t_i and t_j of B_Z are called *Z-compatible* when $Q_{s_i}(u) \cap P_{t_i}$ and $Q_{s_k}(u) \cap P_{t_k}$ are homotopic in $R(t_i, t_k)$ for every $t_k \in B_Z$ with $t_i < t_k \leq t_j$.

Because our family f satisfies the conclusion of Theorem 5.9, Lemma 5.4 has the following consequence: if t_i and t_j are u-compatible for any u, then P_{t_i} and P_{t_j} are both V-cored, or both W-cored, or both bilongitudinal. The next proposition is also immediate from Lemma 5.4.

Proposition 5.9. *Suppose that $t_i, t_j, t_k \in B_Z$. Then at parameters in Z, Q_{s_k} can meet both P_{t_i} and P_{t_j} only if t_i and t_j are Z-compatible.*

For a simplex Δ, write $B_\Delta = \{b_1, \ldots, b_m\}$ with each $b_i < b_{i+1}$, and for each $i \le m$ define a_i to be the s_j for which $b_i = t_j$. Decompose B_Δ into maximal Δ-compatible blocks $C_1 = \{b_1, b_2, \ldots, b_{\ell_1}\}$, $C_2 = \{b_{\ell_1+1}, \ldots, b_{\ell_2}\}, \ldots, C_r = \{b_{\ell_{r-1}+1}, \ldots, b_{\ell_r}\}$, with $\ell_r = m$. Since the blocks are maximal, Proposition 5.9 shows that Q_{a_i} is disjoint from P_{b_j} if b_i and b_j are not in the same block. In steps 3 through 6, this disjointness will ensure that isotopies of these Q_{a_i} do not disturb the results of previous work.

Note that if b_i and b_j lie in the same block, then either both P_{b_i} and P_{b_j} are V-cored, or both are W-cored, or both are bilongitudinal. Thus we can speak of V-cored blocks, and so on.

When δ is a face of Δ, $B_\Delta \subseteq B_\delta$. Therefore if b_i and b_j in B_Δ are δ-compatible, then they are Δ-compatible. So for each block C of B_δ, $C \cap B_\Delta$ is contained in a block of B_Δ. However, levels that are not compatible in B_δ may become compatible in B_Δ, since the t_i for intervening levels in B_δ may fail to be in B_Δ. Typically, the intersections of blocks of B_δ with B_Δ will combine into larger blocks in B_Δ.

We should emphasize that during steps 1 through 6, the blocks of B_Z, and whether a level P_{t_i} is V-cored, W-cored, or bilongitudinal, are defined with respect to the original configuration, not the new positioning after the procedure begins. Indeed, after steps 1 and 2, many of the Q_{s_i} will be disjoint from their P_{t_i}.

We now fill in the details of these procedures.

Step 1: Push the Q_{s_i} that meet P_{t_j} out of V_{t_j}, for all the V-cored P_{t_j}, at all parameters in $U(t_j)$.

We perform this in order of increasing t_j for the V-cored image levels. Begin with t_1. If Q_{s_1} is W-cored or bilongitudinal, do nothing. Suppose it is V-cored. Then for each u in $U(t_1)$, the $Q_{s_j}(u)$ that meet P_{t_1} intersect V_{t_1} in a union of incompressible uniquely boundary-parallel annuli. Since any such Q_{s_j} are transverse to P_{t_1} at each point of $U(t_j)$, the set of intersection annuli $Q_{s_j} \cap V_{t_1}$ falls into finitely many isotopic families, with each family a copy of the connected set $U(t_j)$. For each j with $U(t_1) \cap U(t_j)$ nonempty, let $\mathscr{A}_j$ be the collection of the annuli $Q_{s_j} \cap V_{t_1}$, over all parameters in $U(t_j)$, and let $\mathscr{A}$ be the union of these $\mathscr{A}_j$. The nonempty intersection of $U(t_1)$ and $U(t_j)$ ensures that the loops of $Q_{s_j} \cap P_{t_1}$ and $Q_{s_1} \cap P_{t_1}$ are all in the same isotopy class in P_{t_1}.

One might hope to push these families of annuli out of V_{t_1} one at a time, beginning with an outermost one, but an outermost family might not exist. There could be a sequence $U(t_{j_1}), \ldots, U(t_{j_k})$ such that $U(t_{j_i}) \cap U(t_{j_{i+1}})$ is nonempty for each i, $U(t_{j_k}) \cap U(t_{j_1})$ is nonempty, and for some parameters u_{j_i} in $U(t_{j_i})$, an annulus $Q_{s_{j_{i+1}}}(u_{j_i}) \cap V_{t_1}$ lies outside one of $Q_{s_{j_i}}(u_{j_i}) \cap V_{t_1}$ for each i, and an annulus of $Q_{s_{j_1}}(u_{j_k}) \cap V_{t_1}$ lies outside one of $Q_{s_{j_k}}(u_{j_k}) \cap V_{t_1}$. Since an outermost family might not exist, we will need to utilize the method of Hatcher as in the proof of Theorem 5.9, but only a simple version of it.

Shrink the U_i slightly, obtaining a new open cover by sets U_i' with $\overline{U_i'} \subset U_i$. We will use a function $\Psi\colon \mathscr{A} \to (0,2)$, so that at each parameter u, the restriction ψ_u of Ψ to the annuli at that parameter has the property that $\psi_u(A_1) < \psi_u(A_2)$ whenever $A_1, A_2 \in \mathscr{A}_i$ and A_1 lies in the region of parallelism between A_2 and ∂V_{t_1}. Moreover, we will have $\psi_u(A) < 1$ whenever $A \in \mathscr{A}_i$ and $u \in \overline{U_i'}$, while $\psi_u(A) > 1$ for u near the boundary of U_i. We construct Ψ by letting $\Psi_0(A)$ be the volume of the region of parallelism between A and an annulus in ∂V_{t_1} (assuming that the volume of L has been normalized to 1 to ensure that $\Psi_0(A) < 1$), then adding on auxiliary values $\alpha_i(u)$ as in the proof of Theorem 5.9.

Form the union $G_0 \subset S^d \times (0,2)$ of the $(u, \psi_u(A))$ as in the proof of Theorem 5.9, and thicken each of its sheets as was done there, obtaining an interval for each parameter. These intervals tell the supports of the isotopies that push the annuli of $Q_{s_j} \cap V_{t_1}$ out of V_{t_1}. If two sheets of $\mathscr{A}$ cross in $S^d \times (0,2)$, then the corresponding regions of parallelism have the same volume, so must be disjoint and the isotopies can be performed simultaneously without interference. At each individual parameter u, each annulus is outermost during the time it is being pushed out of V_{t_1}, but the times need to be different since there may be no outermost family.

After the process is completed, Q_{s_j} will lie outside of V_{t_1} at all parameters in $\overline{U(t_j)'}$, whenever $U(t_j)$ had nonempty intersection with $U(t_1)$. Replacing each $U(t_j)$ by $U(t_j)'$, we have Q_{s_j} pushed out of V_{t_1} at all parameters in these $U(t_j)$. Moreover, Lemma 5.3(2) shows that V_{t_1} is concentric in either X_{s_1} or Y_{s_1} at all parameters in $U(t_1)$.

Some of the Q_{s_k} for which $U(t_k)$ did not meet $U(t_1)$ may be moved by the isotopies of the Q_{s_j} at parameters in $U(t_j) \cap U(t_k)$. The condition that these Q_{s_k} meet P_{t_1} transversely may be lost, but this will not matter, because these intersections never matter when $U(t_k)$ does not meet $U(t_1)$.

Now consider t_2. Again, we do nothing if Q_{s_2} is W-cored or bilongitudinal, so suppose that it is V-cored. Use the Hatcher process as before, to push annuli in the Q_{s_j} out of V_{t_2}, when Q_{s_j} meets P_{t_2} and $U(t_j)$ meets $U(t_2)$. Notice that these Q_{s_j} cannot meet V_{t_1} at parameters in $U(t_1)$. For if t_2 is not u-compatible with t_1 at some parameters in $U(t_1)$, then (by Lemma 5.4) Q_{s_j} cannot meet both P_{t_2} and P_{t_1}, while if it is u-compatible at some parameter in $U(t_1)$, then it has already been pushed out of V_{t_1}. And V_{t_1} cannot lie in any of the regions of parallelism for the pushouts from V_{t_2}, since the intersection circles of the Q_{s_j} with P_{t_2} are not longitudes in V_{t_2}.

After these pushouts are completed, if $i = 1$ or $i = 2$ and Q_{s_i} was V-cored, then V_{t_i} is concentric in either X_{s_i} or Y_{s_i} at all parameters in U_i.

We continue working up the increasing t_i in this way. At the end of this process, V_{t_i} is concentric in either X_{s_i} or Y_{s_i} for all i such that Q_{s_i} was V-cored, and at all parameters in U_i. For Q_{s_i} that were W-cored or bilongitudinal, the intersections $Q_{s_i} \cap P_{t_i}$ have not been disturbed at parameters in U_i. We have not introduced any

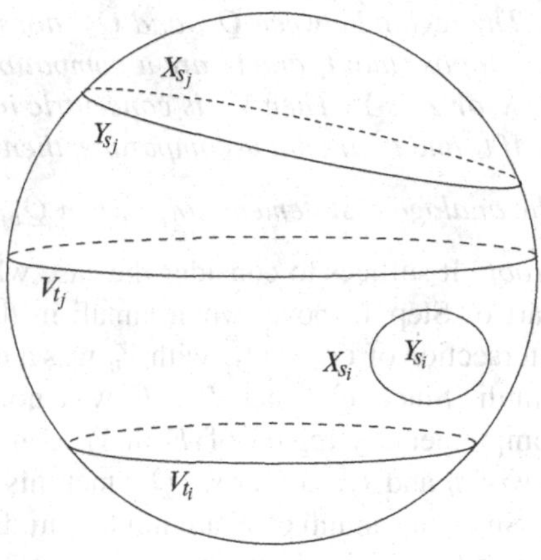

Fig. 5.14 Hypothetical inconsistent nesting: $V_{t_i} \subset X_{s_i}$ and $V_{t_j} \subset Y_{s_j}$

new intersections of Q_{s_i} with P_{t_j}, so we still have the property that at any parameter u in $U_i \cap U(t_j)$, Q_{s_j} can meet P_{t_i} only if t_i and t_j were originally u-compatible.

Step 2: Push the Q_{s_i} that meet P_{t_j} out of W_{t_j} for all the Q_{s_j} that are W-cored, at all parameters in $U(t_j)$.

The entire process is repeated with W-cored levels, except that we start with t_m and proceed in order of decreasing t_i. Each W-cored Q_{s_i} is pushed out of W_{t_i}, and at the end of the process W_{t_i} is concentric in either X_{s_i} or Y_{s_i} at all parameters in U_i, whenever Q_{s_i} was W-cored. No intersection of a Q_{s_j} with a V-cored or bilongitudinal level P_{t_i} is changed at any parameter in U_i.

For the remaining steps, we fix a triangulation of S^d with mesh smaller than a Lebesgue number for $\{U_i\}$, which will ensure that B_Δ is nonempty for every simplex Δ. We will no longer proceed up or down all t_i-levels, working on the sets U_i, but instead will work inductively up the skeleta of the triangulation. Recall that each B_Δ is decomposed into blocks, according to the original intersections of the Q_{s_i} and P_{t_i} before steps 1 and 2 were performed.

Step 3: Push the Q_{s_i} that were originally cored so that each one equals some level torus.

We will proceed inductively up the skeleta of the triangulation, moving cored Q_{s_i} to level tori, without changing $Q_{s_k} \cap P_{s_k}$ for the bilongitudinal Q_{s_k}. We want to use the fact that V_{t_i} (or W_{t_i}) is concentric with X_{s_i} or Y_{s_i} to push Q_{s_i} onto a level torus, but when moving multiple levels at a given parameter, there is a consistency condition needed. As shown in Fig. 5.14, it might happen that V_{t_i} is concentric in X_{s_i} while V_{t_j} is concentric in Y_{s_j}. Then, we might not be able to push Q_{s_i} and Q_{s_j} onto level tori without disrupting other levels. The following lemma rules out this bad configuration.

Lemma 5.18. *Suppose, after steps 1 and 2 have been completed, that $u \in U_i \cap U(t_j)$, $t_i < t_j$, and that Q_{s_i} is V-cored.*

1. *The region between* Q_{s_i} *and* Q_{s_j} *does not contain a core circle of* V_{t_i}.
2. *Suppose that* t_i *and* t_j *are u-compatible, and* V_{t_i} *is concentric in* Z_{s_i} *where* Z *is* X *or* Z *is* Y. *Then* V_{t_j} *is concentric in* Z_{s_j}.
3. *If* t_i *and* t_j *are not u-compatible, then* Q_{s_i} *is parallel to* P_{t_i} *in* $R(t_i, t_j)$.

The analogous statement holds when Q_{s_j} *is* W*-cored and* W_{t_j} *is concentric in* Z_{s_j}.

Proof. It suffices to consider the case when Q_{s_i} is V-cored. In the situation at the start of Step 1 above, when annuli in the Q_{s_k} were being pushed out of V_{t_i}, the intersection of $Q_{s_i} \cup Q_{s_j}$ with V_{t_i} was a union F of incompressible nonlongitudinal annuli. Since Q_{s_i} met P_{t_i}, F was nonempty. By Proposition 5.3, exactly one complementary region of F in V_{t_i} contained a core circle C of V_{t_i}. For at least one of s_i and s_j, say for s_k, Q_{s_k} met this complementary region.

Since the annuli of F are nonlongitudinal, there is an embedded circle C' in Q_{s_k} that is homotopic in the core region to a proper multiple of C. If C were in the region $R = f_u(R(s_i, s_j))$ between Q_{s_i} and Q_{s_j}, then the embedded circle C' in ∂R would be a proper multiple in $\pi_1(R)$, which is impossible since R is homeomorphic to $S^1 \times S^1 \times I$. This proves (1).

Assume that t_i and t_j are u-compatible and suppose that $V_{t_i} \subset X_{s_i}$ and $V_{t_j} \subset Y_{s_j}$. Then C is contained in $X_{s_i} \cap Y_{s_j}$, forcing $s_i > s_j$ and C in the region between Q_{s_i} and Q_{s_j}, contradicting (1). The case of $V_{t_i} \subset Y_{s_i}$ and $V_{t_j} \subset X_{s_j}$ is similar, so (2) holds.

For (3), if t_i and t_j are not u-compatible, then Q_{s_i} was initially disjoint from P_{t_j}, and hence is disjoint after steps 1 and 2. By Lemma 5.3(2), V_{t_i} is concentric in X_{s_i} or Y_{s_i} after steps 1 and 2, and part (3) follows. □

It will be convenient to extend our previous notation $R(s,t)$ for the closure of the region between P_s and P_t, by putting $R(0,t) = V_t$, $R(t,1) = W_t$, and $R(0,1) = L$.

We will now define target regions. The isotopies that we will use in the rest of our process will only change values within a single target region, ensuring that the necessary positioning of the Q_{s_i} is retained. Let Δ be a simplex of the triangulation, and recall the decomposition of $B_\Delta = \{b_1, \ldots, b_m\}$ into maximal Δ-compatible blocks $C_1 = \{b_1, b_2, \ldots, b_{\ell_1}\}$, $C_2 = \{b_{\ell_1+1}, \ldots, b_{\ell_2}\}, \ldots, C_r = \{b_{\ell_{r-1}+1}, \ldots, b_{\ell_r}\}$. Define the *target region* of a block C_n to be the submanifold $T_\Delta(C_n)$ of L defined as follows. Put $\ell_0 = 0$, $b_0 = 0$, and $b_{\ell_r+1} = 1$.

1. If C_n is V-cored, then $T_\Delta(C_n) = R(b_{\ell_{n-1}+1}, b_{\ell_n+1})$.
2. If C_n is W-cored, then $T_\Delta(C_n) = R(b_{\ell_{n-1}}, b_{\ell_n})$.
3. If C_n is bilongitudinal, then $T_\Delta(C_n) = R(b_{\ell_{n-1}}, b_{\ell_n+1})$.

We remark that $T_\Delta(C_n)$ is all of L when B_Δ consists of a single bilongitudinal block, otherwise is of the form V_t when $n = 1$ and C_1 is W-cored or bilongitudinal and of the form W_t when $n = r$ and C_n is V-cored or bilongitudinal, and in all other cases it is a region $R(s,t)$ diffeomorphic to $S^1 \times S^1 \times I$.

As noted in the next lemma, the interior of the target region of a block contains the Q_{a_i} for the b_i in the block, at this point of our argument.

Lemma 5.19. *Target regions satisfy the following.*

1. *If $b_i \in C_n$ and $u \in \Delta$, then $Q_{a_i}(u)$ is in the interior of $T_\Delta(C_n)$.*
2. *If δ is a face of Δ, and $C_1', \ldots, C_{r'}'$ are the blocks of B_δ, then for each i, there exists a j such that $T_\delta(C_i') \subseteq T_\Delta(C_j)$.*

Proof. Property (1) is a consequence of Proposition 5.9 and the fact that Steps 1 and 2 do not create new intersections of the $Q_{s_i}(u)$ with the P_{t_j}. For part (2), the proof is direct from the definitions, dividing into various subcases. □

Target regions can overlap in the following ways: the target region for a V-cored block C_n will overlap the target region of a succeeding W-cored block C_{n+1}, and the target region of a bilongitudinal block will overlap the target region of a preceding V-cored block or of a succeeding W-cored block (note that by Lemma 5.13, successive blocks cannot both be bilongitudinal). The latter cause no difficulties, but the conjunctions of a V-cored block and a succeeding W-cored block will necessitate some care during the ensuing argument.

We can now begin the process that will complete Step 3. We will start at the parameters that are vertices of the triangulation and move the Q_{a_i} for each V-cored or W-cored block to be level, that is, so that each $Q_{a_i}(u)$ equals some P_t. The isotopies will be fixed on each P_{b_i} for which Q_{a_i} is bilongitudinal, and these unchanged $Q_{a_i} \cap P_{b_i}$ will be used to work with the bilongitudinal levels in a later step. For each cored block, the isotopy that levels the Q_{a_i} will move points only in the interior of the target region of the block. As we move to higher-dimensional simplices, the Q_{a_i} will already be level at parameters on the boundary, and the deformation will be fixed at those parameters. Each deformation for the parameters in a simplex δ_0 of dimension less than d must be extended to a deformation of f. The extension will change an f_u only when u is in the open star of δ_0, by a deformation that performs some initial portion of the deformation of f_{u_0} at a parameter u_0 of δ_0—the parameter that is the δ_0-coordinate of u when the simplex that contains it is written as a join $\delta_0 * \delta_1$ (details will be given below). We will see that because the target regions can overlap, the deformation of an f_u might not preserve all target regions, but enough positioning of the image levels Q_{a_i} will be retained to continue the inductive process.

Fix a vertex δ_0 of the triangulation, and consider the first block C_1 of B_{δ_0}. If it is bilongitudinal, we do nothing. Suppose that it is V-cored. All of the $Q_{a_1}, \ldots, Q_{a_{\ell_1}}$ lie in the interior of the target region $T_{\delta_0}(C_1)$. Lemma 5.18(2) shows that for either $Z = X$ or $Z = Y$, V_{b_i} is concentric in Z_{a_i} for $b_i \in C_1$. We claim that there is an isotopy, supported on $T_{\delta_0}(C_1)$, that moves each Q_{a_i} to be level. If C_1 is the only block, then $T_{\delta_0}(C_1) = L$ and the isotopy exists by the definition of concentric. If there is a second block, then Lemma 5.18(3) shows that the Q_{a_i} for $b_i \in C_1$ are parallel to P_{b_1} in $T_{\delta_0}(C_1) = R(b_1, b_{\ell_1+1})$, and again the isotopy exists. After performing the isotopy, we may assume that the $Q_{a_i}(\delta_0)$ are level.

To extend this deformation of f_{δ_0} to a deformation of the parameterized family f, we regard each simplex Δ of the closed star of δ_0 in the triangulation as the join $\delta_0 * \delta_1$, where δ_1 is the face of Δ spanned by the vertices of Δ other than δ_0. Each point of Δ is uniquely of the form $u = s\delta_0 + (1-s)u_1$ with $u_1 \in \delta_1$. Write the isotopy of f_{δ_0} as $j_t \circ f_{\delta_0}$, with j_0 the identity map of L. Then, at u the isotopy at time t is $j_t \circ f_u$ for $0 \le t \le s$ and $j_s \circ f_u$ for $s \le t \le 1$. For any two simplices containing δ_0, this deformation agrees on their intersection, so it defines a deformation of f.

The target region $T_{\delta_0}(C_1)$ will overlap $T_{\delta_0}(C_2)$ if C_2 is bilongitudinal or W-cored. When C_2 is bilongitudinal, this does not affect any of our necessary positioning. If it is W-cored, then Q_{a_i} with $b_i \in C_2$ may be moved into $T_{\delta_0}(C_1)$. At δ_0, such Q_{a_i} can end up somewhere between the now-level $Q_{a_{\ell_1}}$ and $P_{b_{\ell_2}}$, and at other parameters in the star of δ_0 they will lie somewhere in $R(b_1, b_{\ell_2})$. This will require only a bit of attention in the later argument.

In case C_1 was W-cored, we use Lemmas 5.18(2) and 5.17, producing a deformation of f_{δ_0} supported on the interior of the solid torus $T_\Delta(C_1) = V_{b_{\ell_1}}$, which does not meet any other target region. This is extended to a deformation of f just as before.

We move on to consider C_2 in analogous fashion, doing nothing if C_2 is bilongitudinal, and moving the Q_{a_i} to be level at the parameter δ_0. If C_1 was V-cored and C_2 is W-cored, then instead of the initial target region $T_{\delta_0}(C_2)$ we must use the region between the now-level $Q_{a_{\ell_1}}(u)$ and $P_{b_{\ell_2}}$, but otherwise the argument is the same. Proceed in the same way through the remaining blocks C_n of B_{δ_0}, ending with all the cored $Q_{a_i}(u_0)$ moved to be level. This process for u_0 is repeated for each 0-simplex of the triangulation.

Now, consider a simplex δ of positive dimension. Inductively, we may assume that at each u in $\partial\delta$, each cored Q_{a_i} has been moved to a level torus, and $Q_{a_i} \cap Pb_i$ is unchanged for each bilongitudinal Q_{a_i}. Moreover, if a_i is contained in a cored block C_j, then Q_{a_i} lies in the corresponding target region $T_\delta(C_j)$, or else lies in the union of the target regions for a V-cored block and a succeeding W-cored block.

We apply Lemma 5.17 to each cored block of B_δ, sequentially up the cored blocks. We obtain a sequence of deformations of f on δ, constant at parameters in $\partial\delta$. There is no interference between different blocks, except when a W-cored block C_{n+1} succeeds a V-cored block C_n. First, the Q_{a_i} for the V-cored block are moved to be level. Then, at each parameter in δ, the $Q_{a_i}(u)$ for the W-cored block lie between the now-level $Q_{a_{\ell_n}}(u)$ and $P_{b_{\ell_{n+1}}}$. We regard the union of these regions over the parameters of δ as a product $\delta \times S^1 \times S^1 \times \mathrm{I}$, and apply Lemma 5.17. Thus the isotopy that levels the Q_{a_i} from the W-cored block need not move any of the Q_{a_i} from the V-cored block. In other cases, the successive isotopies take place in disjoint regions. To extend this to a deformation of f, we adapt the join method from above (of course when δ is d-dimensional, no extension is necessary). Regard each simplex Δ of the closed star of δ in the triangulation as the join $\delta * \delta_1$, where δ_1 is the face of Δ spanned by the vertices of Δ not in δ. Each point of Δ is uniquely of the form $u = su_0 + (1-s)u_1$ with $u_0 \in \delta$ and $u_1 \in \delta_1$. Write the isotopy of f_{u_0} as $j_t \circ f_{u_0}$, with j_0 the identity map of L. Then, at u the isotopy at time t is $j_t \circ f_u$ for $0 \le t \le s$ and $j_s \circ f_u$ for $s \le t \le 1$. For any two simplices containing δ, this deformation agrees on their intersection, so it defines a deformation of f.

At the completion of this process, each cored Q_{s_i} is level at all parameters in Δ, whenever $\Delta \subset U_i$. The bilongitudinal Q_{s_i} may have been moved around some, but their intersections $Q_{s_i} \cap P_{t_i}$ will not be altered at parameters for which $t_i \in B_\Delta$ since these intersections will not lie in the interior of any target region for a cored level.

Step 4: Push all cored Q_{s_i} to be vertical, that is, make each image of a fiber of P_{s_i} a fiber in L.

Again we work our way up the simplices of the triangulation. Start at a 0-simplex δ_0. Each cored $Q_{a_i}(\delta_0)$ for $b_i \in B_{\delta_0}$ is now level. By Lemma 5.14, the image fibers in $Q_{a_i}(\delta_0)$ are isotopic in that level torus to fibers of L. Using Lemma 5.16, there is an isotopy of f_{δ_0} that preserves the level tori and makes $Q_{a_i}(\delta_0)$ vertical. This isotopy can be chosen to fix all points in other $Q_{a_j}(\delta_0)$, and is extended to a deformation of f by using the method of Step 3. We work our way up the skeleta; if $\delta \subset U(b_i)$, then for every u in δ, each $Q_{a_i}(u)$ is level torus, and at parameters $u \in \partial\delta$, $Q_{a_i}(u)$ is vertical. Using Lemma 5.16, we make the $Q_{a_i}(u)$ vertical at all $u \in \delta$, and extend to a deformation of f as before. We repeat this for all levels of cored blocks.

Step 5: Push all bilongitudinal Q_{s_i} to be vertical.

Now, we examine the bilongitudinal levels. For a bilongitudinal level Q_{a_i} at a vertex δ_0, Corollary 5.1 shows that the intersection circles are longitudes for X_{a_i} and Y_{a_i}. Lemma 5.13 then shows that the circles of $Q_{a_i} \cap P_{b_j}$ are isotopic in Q_{a_i} and in P_{b_j} to fibers. First, use Lemma 4.5 to find an isotopy preserving levels, such that postcomposing f_{δ_0} by the isotopy makes the intersection circles fibers of the P_{b_j}. Then, use Lemma 4.5 applied to $f_{\delta_0}^{-1}$ to find an isotopy preserving levels of the domain, such that precomposing f_{δ_0} by the isotopy makes the intersection circles the images of fibers of P_{s_i}. After this process has been completed for the bilongitudinal Q_{a_i}, the inverse image (in their union $\cup Q_{a_i}$) of each region $R(b_j, b_{j+1})$ with b_j or b_{j+1} in a bilongitudinal block is a collection of fibered annuli which map into $R(b_j, b_{j+1})$ by embeddings that are fiber-preserving on their boundaries. We use Lemma 4.6 to find an isotopy that makes the Q_{a_i} vertical. Again, we extend to a deformation of f and work our way up the skeleta, to assume that $Q_{s_i}(u)$ is vertical whenever $u \in \Delta$ and $\Delta \subset U_i$.

Step 6: Make f fiber-preserving on the complementary $S^1 \times S^1 \times \mathrm{I}$ or solid tori of the P_{s_i}-levels.

We work our way up the skeleta one last time, using Lemma 5.15 to make f fiber-preserving on the complementary $S^1 \times S^1 \times \mathrm{I}$ or solid tori of the P_{a_i}.

There is an annoying technical problem that arises in this step. At each parameter, the deformations that make f_u fiber-preserving on the $S^1 \times S^1 \times \mathrm{I}$ are fixed on the boundaries of these submanifolds, but the extended diffeomorphisms may have to move points on the other side of the frontier. Thus, a region where f_u was already fiber-preserving may be changed to make f_u no longer fiber-preserving there. One fix for this is as follows. We can arrange that the final f has all f_u fiber-preserving except on small product neighborhoods of a finite set of levels at each parameter. Then by removing a neighborhood of the singular circles and their images, we can regard f as a parameterized family of diffeomorphisms of $S^1 \times S^1 \times \mathrm{I}$ that is fiber-preserving on a neighborhood of the boundary at each parameter. Then we apply the following version of Theorem 3.13:

Theorem 5.11. *Suppose that Σ is a Seifert-fibered three-manifold with boundary and $g\colon \Sigma \times W \to \Sigma$ is a parameterized family of diffeomorphisms, with W compact, such that each g_u is fiber-preserving on a neighborhood of $\partial\Sigma$. Then there is a deformation of g, relative to $U \times W$ for some open neighborhood U of $\partial\Sigma$ in Σ, to a family of fiber-preserving diffeomorphisms.*

To prove this, we know from Theorem 3.13 that there is some deformation from g to a family h of fiber-preserving diffeomorphisms. Since the inclusion $\mathrm{diff}_f(\partial\Sigma) \to \mathrm{diff}(\partial\Sigma)$ is a homotopy equivalence, the restriction of this deformation to $\partial\Sigma$ can be assumed to be fiber-preserving at all times. Choosing a collar $\partial\Sigma \times \mathrm{I}$ in which each $\Sigma \times \{t\}$ is a union of fibers, we may use uniqueness of collars to change the deformation to be fiber-preserving on the collar. Now, by performing less and less of the deformation as one moves toward $\partial\Sigma$, obtain a new deformation from g to a fiber-preserving family h' such that $h' = h$ outside $\partial\Sigma \times \mathrm{I}$, but $h = g$ on $\partial\Sigma \times [0, 1/2]$. □

5.13 Parameters in D^d

Regard D^d as the unit ball in d-dimensional Euclidean space, with boundary the unit sphere S^{d-1}. As mentioned in Sect. 5.2, to prove that $\mathrm{diff}_f(L) \to \mathrm{diff}(L)$ is a homotopy equivalence, we actually need to work with a family of diffeomorphisms f of L parameterized by D^d, $d \geq 1$, for which $f(u)$ is fiber-preserving whenever u lies in the boundary S^{d-1}. We must deform f so that each $f(u)$ is fiber-preserving, by a deformation that keeps $f(u)$ fiber-preserving at all times when $u \in S^{d-1}$.

We now present a trick that allows us to gain good control of what happens on S^{d-1}. The Hopf fibering we are using on L can be described as a Seifert fibering of L over the round 2-sphere S, in such a way that each isometry of L projects to an isometry of S. For the cases when $q = 1$, the round sphere is the actual quotient orbifold, and when $1 < q$, the quotient orbifold has two cone points but the only induced isometries are rotations fixing those cone points. (Section 4.4 above details this description for the manifolds considered in Chap. 4, full details of all cases are in [46].) By conjugating $\pi_1(L)$ in SO(4), we may assume that the singular fibers, when $q > 1$, are the inverse images of the poles. We choose our sweepout so that the level tori are the inverse images of latitude circles. Denote by p_t the latitude circle that is the image of the level torus P_t.

There is an isotopy J_t with J_0 the identity map of L and each J_t fiber-preserving, so that the images of the level tori P_s under J_1 project to circles in the 2-sphere as indicated in Fig. 5.15. Denote the image of $J_1(P_s)$ in S by q_s. Their key property is that when moved by any orthogonal rotation of S, each p_t meets the image of some q_s transversely in two or four points.

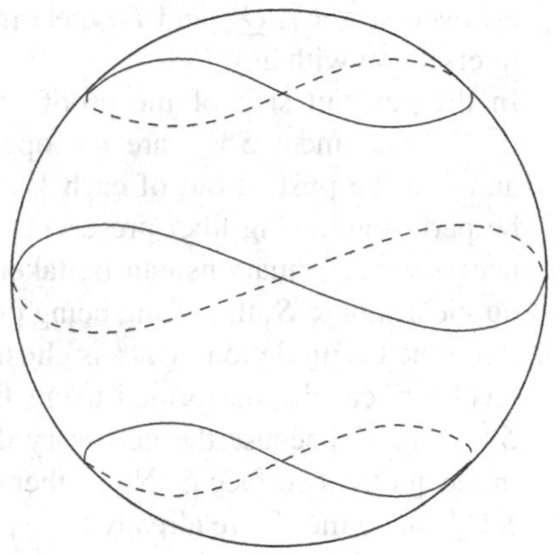

Fig. 5.15 Projections of the $J_1(P_t)$ into the 2-sphere

Using Theorem 5.1, we may assume that f_u is actually an isometry of L for each $u \in S^{d-1}$. Denote the isometry that f_u induces on S by $\overline{f_u}$. Now, deform the entire family f by precomposing each f_u with J_t. At points in S^{d-1}, each $f_u \circ J_t$ is fiber-preserving, so this is an allowable deformation of f. At the end of the deformation, for each $u \in S^{d-1}$, $f_u \circ J_1(P_s)$ is a fibered torus Q_s that projects to $\overline{f_u}(q_s)$. Since $\overline{f_u}$ is an isometry of S, it follows that for any latitude circle p_t, some $\overline{f_u}(q_s)$ meets p_t transversely, in either two or four points. So P_t and this Q_s meet transversely in either two or four circles which are fibers of L. In particular, they are in very good position. We call such a pair P_t and Q_s at u an *instant pair*.

Cover S^{d-1} by finitely many open sets Z_i' such that for each i, there is an (x_i, y_i) such that Q_{x_i} and P_{y_i} are an instant pair at every point of $\overline{Z_i'}$. We may assume that there are open sets Z_i in D^d such that $\overline{Z_i} \cap S^{d-1} = \overline{Z_i'}$ and Q_{x_i} and P_{y_i} meet in very good position at each point of $\overline{Z_i}$. For any sufficiently small deformation of f, both Q_{x_i} and P_{y_i} will still meet in very good position at all points of $\overline{Z_i}$. Let V be a neighborhood of S^{d-1} in D^d such that $\overline{V}$ is contained in the union of the Z_i.

Now, we apply to D^d the entire process used for the case when the parameters lie in S^d, using appropriate fiber-preserving deformations at parameters in S^{d-1}. Here are the steps:

1. By Theorem 5.3, there are arbitrarily small deformations of f that put it in general position with respect to the sweepout. Select the deformation sufficiently small so that the Q_{x_i} and P_{y_i} still meet in very good position at every point of $\overline{Z_i}$. Within V, we taper the deformation off to the identity, so that no change has taken place at parameters in S^{d-1}. At every parameter, either there is already a pair in very good position, or f_u satisfies the conditions (GP1), (GP2), and (GP3) of a general position family.
2. Theorem 5.8 guarantees that at each of the parameters in $D^d - V$, there is a pair Q_s and P_t meeting in good position.
3. Applying Theorem 5.9 to D^d, with S^{d-1} in the role of W_0, we find a deformation of f, fixed on S^{d-1}, and a covering U_i of D^d and associated values s_i so that

for every $u \in U_i$, Q_{s_i} and P_{t_i} meet in very good position, and Q_{s_i} has no discal intersection with any P_{t_j}.

4. In the pushout step of the proof of Theorem 5.10, we may assume that all the U_i that meet S^{d-1} are the open sets Z_i. At parameters u in S^{d-1}, the annuli to be pushed out of each V_{t_i} will be vertical annuli. The pushouts may be performed using fiber-preserving isotopies at these parameters, because the necessary deformations can be taken as lifts of deformations of circles in the quotient sphere S, the lifting being possible by Theorem 3.11.
5. After the triangulation of D^d is chosen, the deformation that move the Q_{s_i} onto level tori can be performed using fiber-preserving isotopies at parameters in S^{d-1}, again because the necessary deformations cover deformations of circles in the quotient surface S. No further deformation will be needed on simplices in S^{d-1}, since the f_u are already fiber-preserving there.

This completes the discussion of the case of parameters in D^d, and the proof of the Smale Conjecture for lens spaces.

References

1. Aneziris, C., Balachandran, A.P., Bourdeau, M., Jo, S., Sorkin, R.D., Ramadas, T.R.: Aspects of spin and statistics in generally covariant theories. Int. J. Mod. Phys. A **4**(20), 5459–5510 (1989)
2. Asano, K.: Homeomorphisms of prism manifolds. Yokohama Math. J. **26**(1), 19–25 (1978)
3. Banyaga, A.: The structure of classical diffeomorphism groups. In: Mathematics and Its Applications, vol. 400. Kluwer Academic, Dordrecht (1997)
4. Bessaga, C., Pelczynski, A.: Selected topics in infinite-dimensional topology. In: Monografie Matematyczne, Tom 58 [Mathematical Monographs, vol. 58]. PWN—Polish Scientific Publishers, Warsaw (1975)
5. Boileau, M., Otal, J.P.: Scindements de Heegaard et groupe des homeotopies des petites varietes de Seifert. Invent. Math. **106**(1), 85–107 (1991)
6. Bonahon, F.: Difféotopies des espaces lenticulaires. Topology **22**(3), 305–314 (1983)
7. Bredon, G.E., Wood, J.W.: Non-orientable surfaces in orientable 3-manifolds. Invent. Math. **7**, 83–110 (1969)
8. Bruce, J.W.: On transversality. Proc. Edinb. Math. Soc. (2) **29**(1), 115–123 (1986)
9. Casson, A.J., Gordon, C.McA.: Reducing Heegaard splittings. Topology Appl. **27**(3), 275–283 (1987)
10. Cerf, J.: Topologie de certains espaces de plongements. Bull. Soc. Math. Fr. **89**, 227–380 (1961)
11. Cerf, J.: Sur les difféomorphismes de la sphere de dimension trois ($\Gamma_4 = 0$). In: Lecture Notes in Mathematics, vol. 53. Springer, Berlin (1968)
12. Charlap, L.S., Vasquez, A.T.: Compact flat riemannian manifolds. III. The group of affinities. Am. J. Math. **95**, 471–494 (1973)
13. Filipkiewicz, R.P.: Isomorphisms between diffeomorphism groups. Ergod. Theor. Dyn. Syst. **2**(2), 159–171 (1982)
14. Friedman, J., Sorkin, R.: Spin 1/2 from gravity. Phys. Rev. Lett. **44**, 1100–1103 (1980)
15. Gabai, D.: The Smale conjecture for hyperbolic 3-manifolds: $\mathrm{Isom}(M^3) \simeq \mathrm{Diff}(M^3)$. J. Differ. Geom. **58**(1), 113–149 (2001)
16. Gibson, C.G., Wirthmuller, K., du Plessis, A.A., Looijenga, E.J.N.: Topological stability of smooth mappings. In: Lecture Notes in Mathematics, vol. 552. Springer, Berlin (1976)
17. Giulini, D.: On the configuration space topology in general relativity. Helv. Phys. Acta **68**(1), 86–111 (1995)
18. Giulini, D., Louko, J.: No-boundary θ sectors in spatially flat quantum cosmology. Phys. Rev. D (3) **46**(10), 4355–4364 (1992)
19. Gramain, A.: Le type d'homotopie du groupe des difféomorphismes d'une surface compacte. Ann. Sci. École Norm. Sup. (4) **6**, 53–66 (1973)

S. Hong et al., *Diffeomorphisms of Elliptic 3-Manifolds*, Lecture Notes in Mathematics 2055, DOI 10.1007/978-3-642-31564-0,

20. Hamilton, R.S.: The inverse function theorem of Nash and Moser. Bull. Am. Math. Soc. (N.S.) **7**(1), 65–222 (1982)
21. Hantsche, W., Wendt, W.: Drei dimensionali Euklidische Raumformen. Math. Ann. **110**, 593–611 (1934)
22. Hatcher, A.E.: Homeomorphisms of sufficiently large P^2-irreducible 3-manifolds. Topology **15**(4), 343–347 (1976)
23. Hatcher, A.E.: On the diffeomorphism group of $S^1 \times S^2$. Proc. Am. Math. Soc. **83**(2), 427–430 (1981)
24. Hatcher, A.E.: A proof of the Smale conjecture, Diff$(S^3) \simeq$ O(4). Ann. Math. (2) **117**(3), 553–607 (1983)
25. Hatcher, A.: On the diffeomorphism group of $S^1 \times S^2$. Revised version posted at http://www.math.cornell.edu/~hatcher/#papers (2003)
26. Hempel, J.: 3-Manifolds. In: Annals of Mathematics Studies, vol. 86. Princeton University Press, Princeton (1976)
27. Henderson, D.W.: Corrections and extensions of two papers about infinite-dimensional manifolds. Gen. Topology Appl. **1**, 321–327 (1971)
28. Henderson, D.W., Schori, R.: Topological classification of infinite dimensional manifolds by homotopy type. Bull. Am. Math. Soc. **76**, 121–124 (1970)
29. Hendriks, H.: La stratification naturelle de l'espace des fonctions différentiables reelles n'est pas la bonne. C. R. Acad. Sci. Paris Ser. A-B **274**, A618–A620 (1972)
30. Isham, C.J.: Topological θ-sectors in canonically quantized gravity. Phys. Lett. B **106**(3), 188–192 (1981)
31. Ivanov, N.V.: Groups of diffeomorphisms of Waldhausen manifolds, Studies in topology, II. Zap. Naučn. Sem. Leningrad. Otdel. Mat. Inst. Steklov. (LOMI) **66**, 172–176, 209 (1976)
32. Ivanov, N.V.: Diffeomorphism groups of Waldhausen manifolds. J. Sov. Math. **12**, 115–118 (1979)
33. Ivanov, N.V.: Homotopies of automorphism spaces of some three-dimensional manifolds. Dokl. Akad. Nauk SSSR **244**(2), 274–277 (1979)
34. Ivanov, N.V.: Corrections: Homotopies of automorphism spaces of some three-dimensional manifolds. Dokl. Akad. Nauk SSSR **249**(6), 1288 (1979)
35. Ivanov, N.V.: Homotopy of spaces of diffeomorphisms of some three-dimensional manifolds. Zap. Nauchn. Sem. Leningrad. Otdel. Mat. Inst. Steklov. (LOMI) **122**, 72–103, 164–165 (1982)
36. Ivanov, N.V.: Homotopy of spaces of diffeomorphisms of some three-dimensional manifolds. J. Soviet Math. **26**, 1646–1664 (1984)
37. Jaco, W.: Lectures on three-manifold topology. In: CBMS Regional Conference Series in Mathematics, vol. 43. American Mathematical Society, Providence (1980)
38. Jaco, W., Shalen, P.B.: Seifert fibered spaces in 3-manifolds. Memoir. Am. Math. Soc. **21**(220), viii+192 (1979)
39. Kalliongis, J., McCullough, D.: Isotopies of 3-manifolds. Topology Appl. **71**(3), 227–263 (1996)
40. Karcher, H.: Riemannian center of mass and mollifier smoothing. Comm. Pure Appl. Math. **30**(5), 509–541 (1977)
41. Kobayashi, T., Saeki, O.: The Rubinstein-Scharlemann graphic of a 3-manifold as the discriminant set of a stable map. Pac. J. Math. **195**(1), 101–156 (2000)
42. Kriegl, A., Michor, P.W.: The convenient setting of global analysis. In: Mathematical Surveys and Monographs, vol. 53. American Mathematical Society, Providence (1997)
43. Friedman, J.L., Witt, D.M.: Homotopy is not isotopy for homeomorphisms of 3-manifolds. Topology **25**(1), 35–44 (1986)
44. Lundell, A., Weingram, S.: The Topology of CW Complexes. Van Nostrand Reinhold, Princeton (1969)
45. Mather, J.N.: Stability of C^∞ mappings. III. Finitely determined mapgerms. Inst. Hautes Etudes Sci. Publ. Math. **35**, 279–308 (1968)
46. McCullough, D.: Isometries of elliptic 3-manifolds. J. Lond. Math. Soc. (2) **65**(1), 167–182 (2002)

47. McCullough, D., Soma, T.: The Smale conjecture for Seifert fibered spaces with hyperbolic base orbifold (2010) [ArXiv:1005.5061]
48. Neumann, W., Raymond, F.: Automorphisms of Seifert manifolds (1979). Preprint
49. Orlik, P.: Seifert manifolds. In: Lecture Notes in Mathematics, vol. 291. Springer, Berlin (1972)
50. Orlik, P., Vogt, E., Zieschang, H.: Zur Topologie gefaserter dreidimensionaler Mannigfaltigkeiten. Topology **6**, 49–64 (1967)
51. Palais, R.S.: Local triviality of the restriction map for embeddings. Comment. Math. Helv. **34**, 305–312 (1960)
52. Palais, R.S.: Homotopy theory of infinite dimensional manifolds. Topology **5**, 1–16 (1966)
53. Park, C.Y.: Homotopy groups of automorphism groups of some Seifert fiber spaces. Dissertation at the University of Michigan (1989)
54. Park, C.Y.: On the weak automorphism group of a principal bundle, product case. Kyungpook Math. J. **31**(1), 25–34 (1991)
55. Pitts, J.T., Rubinstein, J.H.: Applications of minimax to minimal surfaces and the topology of 3-manifolds. Miniconference on geometry and partial differential equations, 2 (Canberra, 1986), pp. 137–170. In: Proc. Centre Math. Anal. Austral. Nat. Univ., vol. 12. Austral. Nat. Univ., Canberra (1987)
56. Rubinstein, J.H.: On 3-manifolds that have finite fundamental group and contain Klein bottles. Trans. Am. Math. Soc. **251**, 129–137 (1979)
57. Rubinstein, J.H., Birman, J.S.: One-sided Heegaard splittings and homeotopy groups of some 3-manifolds. Proc. Lond. Math. Soc. (3) **49**(3), 517–536 (1984)
58. Rubinstein, J.H., Scharlemann, M.: Comparing Heegaard splittings of non-Haken 3-manifolds. Topology **35**(4), 1005–1026 (1996)
59. Sakuma, M.: The geometries of spherical Montesinos links. Kobe J. Math. **7**(2), 167–190 (1990)
60. Scott, P.: The geometries of 3-manifolds. Bull. Lond. Math. Soc. **15**(5), 401–487 (1983)
61. Seeley, R.T.: Extension of C^∞ functions defined in a half space. Proc. Am. Math. Soc. **15**, 625–626 (1964)
62. Seifert, H.: Topologie dreidimensionaler gefaserter raume. Acta Math. **60**(1), 147–238 (1933)
63. Sergeraert, F.: Un théoréme de fonctions implicites sur certains espaces de Fréchet et quelques applications. Ann. Sci. École Norm. Sup. (4) **5**, 599–660 (1972)
64. Smale, S.: Diffeomorphisms of the 2-sphere. Proc. Am. Math. Soc. **10**, 621–626 (1959)
65. Sorkin, R.D.: Classical topology and quantum phases: Quantum geons. In: Geometrical and Algebraic Aspects of Nonlinear Field Theory. North-Holland Delta Series, pp. 201–218. North-Holland, Amsterdam (1989)
66. Takens, F.: Characterization of a differentiable structure by its group of diffeomorphisms. Bol. Soc. Brasil. Mat. **10**(1), 17–25 (1979)
67. Tougeron, J.C.: Une généralisation du théoréme des fonctions implicites. C. R. Acad. Sci. Paris Ser. A-B **262**, A487–A489 (1966)
68. Tougeron, J.C.: Idéaux de fonctions différentiables. I. Ann. Inst. Fourier (Grenoble) **18**(fasc. 1), 177–240 (1968)
69. Waldhausen, F.: Eine Klasse von 3-dimensionalen Mannigfaltigkeiten. I, II. (German). Invent. Math. **3**, 308–333 (1967). Ibid: **4**, 87–117 (1967)
70. Waldhausen, F.: Gruppen mit Zentrum und 3-dimensionale Mannigfaltigkeiten. Topology **6**, 505–517 (1967)
71. Waldhausen, F.: On irreducible 3-manifolds which are sufficiently large. Ann. Math. (2) **87**, 56–88 (1968)
72. Wall, C.T.C.: Finite determinacy of smooth map-germs. Bull. Lond. Math. Soc. **13**(6), 481–539 (1981)
73. Witt, D.M.: Symmetry groups of state vectors in canonical quantum gravity. J. Math. Phys. **27**(2), 573–592 (1986)
74. Wolf, J.A.: Spaces of Constant Curvature, 3rd edn. Publish or Perish Inc., Boston (1974)

Index

Mathematical symbols appear separately at the beginning of the index, and are repeated in their lexicographical position later. Page numbers given in **boldface** indicate the location of the main definition of the item, and those in *italics* indicate statements of results.

S. Hong et al., *Diffeomorphisms of Elliptic 3-Manifolds*, Lecture Notes in Mathematics 2055, DOI 10.1007/978-3-642-31564-0,

LECTURE NOTES IN MATHEMATICS Springer

Edited by J.-M. Morel, B. Teissier; P.K. Maini

Editorial Policy (for the publication of monographs)

1. Lecture Notes aim to report new developments in all areas of mathematics and their applications - quickly, informally and at a high level. Mathematical texts analysing new developments in modelling and numerical simulation are welcome.
   Monograph manuscripts should be reasonably self-contained and rounded off. Thus they may, and often will, present not only results of the author but also related work by other people. They may be based on specialised lecture courses. Furthermore, the manuscripts should provide sufficient motivation, examples and applications. This clearly distinguishes Lecture Notes from journal articles or technical reports which normally are very concise. Articles intended for a journal but too long to be accepted by most journals, usually do not have this "lecture notes" character. For similar reasons it is unusual for doctoral theses to be accepted for the Lecture Notes series, though habilitation theses may be appropriate.

2. Manuscripts should be submitted either online at www.editorialmanager.com/lnm to Springer's mathematics editorial in Heidelberg, or to one of the series editors. In general, manuscripts will be sent out to 2 external referees for evaluation. If a decision cannot yet be reached on the basis of the first 2 reports, further referees may be contacted: The author will be informed of this. A final decision to publish can be made only on the basis of the complete manuscript, however a refereeing process leading to a preliminary decision can be based on a pre-final or incomplete manuscript. The strict minimum amount of material that will be considered should include a detailed outline describing the planned contents of each chapter, a bibliography and several sample chapters.
   Authors should be aware that incomplete or insufficiently close to final manuscripts almost always result in longer refereeing times and nevertheless unclear referees' recommendations, making further refereeing of a final draft necessary.
   Authors should also be aware that parallel submission of their manuscript to another publisher while under consideration for LNM will in general lead to immediate rejection.

3. Manuscripts should in general be submitted in English. Final manuscripts should contain at least 100 pages of mathematical text and should always include
   - a table of contents;
   - an informative introduction, with adequate motivation and perhaps some historical remarks: it should be accessible to a reader not intimately familiar with the topic treated;
   - a subject index: as a rule this is genuinely helpful for the reader.

   For evaluation purposes, manuscripts may be submitted in print or electronic form (print form is still preferred by most referees), in the latter case preferably as pdf- or zipped psfiles. Lecture Notes volumes are, as a rule, printed digitally from the authors' files. To ensure best results, authors are asked to use the LaTeX2e style files available from Springer's web-server at:

   ftp://ftp.springer.de/pub/tex/latex/svmonot1/ (for monographs) and
   ftp://ftp.springer.de/pub/tex/latex/svmultt1/ (for summer schools/tutorials).

Additional technical instructions, if necessary, are available on request from lnm@springer.com.

4. Careful preparation of the manuscripts will help keep production time short besides ensuring satisfactory appearance of the finished book in print and online. After acceptance of the manuscript authors will be asked to prepare the final LaTeX source files and also the corresponding dvi-, pdf- or zipped ps-file. The LaTeX source files are essential for producing the full-text online version of the book (see http://www.springerlink.com/openurl.asp?genre=journal&issn=0075-8434 for the existing online volumes of LNM). The actual production of a Lecture Notes volume takes approximately 12 weeks.

5. Authors receive a total of 50 free copies of their volume, but no royalties. They are entitled to a discount of 33.3 % on the price of Springer books purchased for their personal use, if ordering directly from Springer.

6. Commitment to publish is made by letter of intent rather than by signing a formal contract. Springer-Verlag secures the copyright for each volume. Authors are free to reuse material contained in their LNM volumes in later publications: a brief written (or e-mail) request for formal permission is sufficient.

Addresses:

Professor J.-M. Morel, CMLA,
École Normale Supérieure de Cachan,
61 Avenue du Président Wilson, 94235 Cachan Cedex, France
E-mail: morel@cmla.ens-cachan.fr

Professor B. Teissier, Institut Mathématique de Jussieu,
UMR 7586 du CNRS, Équipe "Géométrie et Dynamique",
175 rue du Chevaleret
75013 Paris, France
E-mail: teissier@math.jussieu.fr

For the "Mathematical Biosciences Subseries" of LNM:

Professor P. K. Maini, Center for Mathematical Biology,
Mathematical Institute, 24-29 St Giles,
Oxford OX1 3LP, UK
E-mail : maini@maths.ox.ac.uk

Springer, Mathematics Editorial, Tiergartenstr. 17,
69121 Heidelberg, Germany,
Tel.: +49 (6221) 4876-8259

Fax: +49 (6221) 4876-8259
E-mail: lnm@springer.com